T0146199

Functional Inference in Paleoanthropology

Functional Inference in Paleoanthropology

Theory and Practice

David J. Daegling

Johns Hopkins University Press
Baltimore

© 2022 Johns Hopkins University Press
All rights reserved. Published 2022
Printed in the United States of America on acid-free paper
9 8 7 6 5 4 3 2 1

Johns Hopkins University Press
2715 North Charles Street
Baltimore, Maryland 21218-4363
www.press.jhu.edu

Library of Congress Cataloging-in-Publication Data

Names: Daegling, David J., 1960– author.
Title: Functional inference in paleoanthropology : theory and practice /
    David J. Daegling.
Description: Baltimore, Maryland : Johns Hopkins University Press, [2022] |
    Includes bibliographical references and index.
Identifiers: LCCN 2021011252 | ISBN 9781421442945 (hardcover) |
    ISBN 9781421442952 (ebook)
Subjects: LCSH: Paleoanthropology.
Classification: LCC GN281 .D34 2022 | DDC 569.9—dc23
LC record available at https://lccn.loc.gov/2021011252

A catalog record for this book is available from the British Library.

*Special discounts are available for bulk purchases of this book. For more
information, please contact Special Sales at specialsales@jh.edu.*

In memory of Mary, Bill, John, and Karl

Contents

Preface

Several years ago, my colleague Peter Ungar invited me to participate in a conference on hominin dietary adaptations and evolution. The motley collection of attendees included anatomists, archaeologists, behavioral ecologists, human biologists, and pop-diet gurus. The ground rules were unusual: participants were instructed to address their particular research specializations in terms of what was known, what was unknown, and what was unknowable. Several participants ignored the directive, but most of us took these instructions seriously. There was considerable variation along what could be characterized as a pessimism–optimism axis regarding the question of whether the paleoanthropological community had the theoretical acumen and methodological means to reconstruct early hominin diets. I occupied the far reaches of the pessimistic end, and on the day of my talk, I argued that even if we could understand the basic functions reflected in fossil anatomy, we had scant hope of deciphering diet from that information. Later that evening a distinguished primatologist diagnosed the error in my thinking, telling me, "You only believe these things because your experiments and comparisons didn't work out." I took this to mean that the proper attitude was that a clear extrapolation from fossil morphology to ecology and behavior was inevitable, given correctness of thought.

Indeed, my "problematic" perspective had arisen because, in my work, I had failed to uncover a neat and tidy connection between adaptation and morphology, at least as far as could be ascertained through the lens of biomechanics. I came to believe my skepticism about what we could reconstruct about human evolution from the fossil record was due to something other than a poor disposition. My reasoning and resulting advocacy of a measured skepticism is the subject of this book. Having received much valid, and sometimes painful, criticism from innumerable colleagues over the years, I harbor no illusions that every pronouncement I express in the following pages will be judged to be correct, or even reasonable. My objective, however, is to further a disciplinary conversation about why functional inference in the fossil record is so vexing when the logic of natural selection suggests that adaptation should be ubiquitously represented there.

I owe a great many thanks to friends and colleagues who have lent assistance in the preparation of this book. Scott McGraw, Matt Ravosa, and Daniel Schmitt read and commented on various chapters; the number of errors found here would be magnified significantly were it not for their taking the time to offer cogent critiques. Henna Bhramdat spent countless hours implementing and debugging code to make the bone growth simulations described in Chapter 6 work properly, and Viviana Toro-Ibacache provided much-needed theoretical and practical guidance with respect to these models. I am grateful to Esther Rodriguez at Johns Hopkins University Press for her assistance with figure formatting and preparation, and to my editor, Tiffany Gasbarrini, for taking a gamble on an unconventional book proposal, advocating on my behalf, and making the whole process (outside of writing) a stress-free experience. Kathleen Capels copyedited the manuscript and repaired several poorly executed turns of phrase. Lastly, I would like to thank my wife Marnie for her unwavering support, both tangible (editing the first and last chapters) and intangible, and my son Zach, for regularly getting me into the surf and away from screens to remind me of what is important.

Functional Inference in Paleoanthropology

Chapter One

Unresolved Problems in Human Evolution

S ometime before four million years ago, groups of primates began walking upright, on two legs, and one of these populations gave rise to us. As there are speedier and more stable ways for primates to move around, we have sought to discover why these primates stood up and walked. In retrospect, it seems as if little needs explaining. We can see how bipedality freed the hands to explore material technology, and we take it for granted that our brains and language inevitably followed from this radical locomotor transition.

The question of how bipedalism began, however, actually remains very much unresolved. There has been no shortage of explanatory offerings, all of which posit the emergence of bipedality as meeting a novel ecological or behavioral need. This reasoning typically invokes an ecological problem that our ancestors cracked, with a cascade of consequent changes that led to our domination of the biosphere. Yet nearly a century and a half after Charles Darwin first posed the question of our locomotor origins, there is no emerging consensus as to what that ecological challenge was.

The easiest explanation for our irresolution is that a dearth of fossil evidence is responsible. The plaint that "we need more fossils" is a near-mandatory caveat in publications dealing with hominin evolution. The fossil hominin inventory, however, has become much richer in recent years. Despite this, our discrimination regarding the selective forces driving hominin adaptation has not improved in proportion to this enhancement of that inventory. Questions concerning how evolutionary processes shaped bipedality, tool manufacture, diet, and language are still hotly contested.

Our inability to resolve questions of adaptation and behavior from the fossil record has less to do with the number of fossils than it does with what we believe to be their information content. Fossils are mineralized casts of the bones of long-deceased individuals who, in nearly all cases, died from something other than an adaptive deficiency of the remains in our possession. That being the case, they would seem to offer a very narrow window into past behavior. As a result, paleoanthropologists wisely draw on information about paleoenvironments to inform functional and adaptive arguments. This top-down perspective is useful, up to a point. It is logical to assume that the morphology of the fossils somehow reflects a successful negotiation of one or more environmental challenges. But that reflection is limited. Environmental variation does not map onto osteological variation with good precision.

There is another perspective, however: a distinct bottom-up means of interpreting fossils that has been underutilized in paleoanthropology. This approach focuses on the fact that fossils were once bone—a living, reactive tissue that responds to environmental stimuli in predictable ways. Bone tunes itself to the physical environment, responding to changes in activity patterns during its growth and development, as well as over evolutionary time. While paleoanthropologists may appreciate this certainty, our operative assumption has been that bone has the ability to refashion itself, materially and structurally, to adapt to its circumstances in a more or less ideal way. An abundance of circumstantial evidence has reassured paleoanthropologists that this assumption is true. But over the last 30 years, a growing body of research in the field of skeletal biology has emerged that suggests the need for a reconsideration of this view. I delve into the implications of this research further on, but, for the moment, suffice it to say that the environment an organism's skeleton "sees" is different, and much simpler, than the environment envisioned by the paleoanthropologist. We will do better at deciphering the adaptive and behavioral significance of fossil morphology if we pay attention to this fact.

Functional morphology, as a subspecialty of evolutionary biology, is a strange bird. The development of the discipline of systematics depended on the prior existence of evolutionary theory (taxonomy existed prior to evolutionary thought, but phylogeny did not), yet functional morphology did quite well prior to the time when evolution started gaining traction. Evolutionary theory is not necessary to figure out how animal behavior is linked to anatomy, and this fact has troubled a generation of morphologists. The mechanics of function are not difficult to discover, provided you can measure the things you need and do some math. Relating those numbers to an evolutionary process is what becomes challenging. This realization has led to offshoots of functional morphology, with new subdisciplines of evolutionary morphology, ecomorphology, and constructional morphology emerging over the past half century. Obviously, describing the study of morphology as "evolutionary" rather than "functional" is a vacuous distinction if research practice and perspective are essentially the same in both areas. Accordingly, Lauder (1981) developed a set of principles in which the study of form was explicitly situated in the context of historical change, where ideas about selection, constraint, and diversification could be sensibly incorporated into an evolutionary framework.

An important and largely unexamined issue in paleoanthropology is whether the ways in which we draw functional inferences (whatever we call that endeavor) are coherent with respect to evolutionary processes. Another way of framing this question is whether the methods used to draw such inferences reflect the broader, underlying theoretical perspectives that define evolutionary biology. The theory behind evolutionary morphology, with respect to the fossil record, is less tangible than the theoretical concerns of systematics. My purpose in this book is to make that theory more explicit and consider whether current methods can intelligibly address the central questions pertaining to adaptation and the reconstruction of behavior in hominin evolution.

In his seminal work, *Adaptation and Natural Selection*, Williams (1966) argued for clarity and parsimony in the application of evolutionary theory. The uncritical invocation of adaptation was his primary target, and he noted that without formal criteria for its recognition, adaptation could be used to explain anything utilitarian in biological systems (his examples included fox feet as adaptations for trampling snow, and flying fish morphology as an adaptation for returning to the sea). So pervasive was this problem that Williams (1966) advocated a new field of inquiry, called teleonomy, which he envisioned as a scientific study specifically examining adaptation. This field was to be situated within broader evolutionary theory, but it would introduce a formalism into ethology and morphology that was heretofore missing. Though there have been significant steps toward such a formalism in the more than 50 years since this proposal was first set out, there remains no such recognized field of inquiry. In some ways, evolutionary morphology has provided the conceptual focus Williams (1966) demanded, but biologists of all stripes still view the definition and recognition of adaptation as problematic.

We all appreciate that morphology is not static, but operationalizing how morphology changes in an evolutionary context is challenging. Ultimately, our goal as paeloanthropologists is to describe how hominins, as organisms, evolved. Since fossils (with a few exceptions) were once bones,[1] I suggest that a consideration of the particulars of how bone tissue evolves is a necessary part of that investigation.

Paleoanthropology, as a discipline, has had no formal consensus on how to draw functional inferences from the fossil record. It employs only an inventory of methods regarded as helpful to that effort. There are three reasons why this might be: (1) there is no need for formal criteria, as the approaches in use are transparent and effective; (2) such formality is avoided, because its invocation would expose the weakness of our methods and the fruitlessness of pursuing such questions; or (3) there is an incongruence between theoretical premises and methodological design that has largely gone unrecognized (i.e., we are gleaning answers from the wrong data). The first of these reasons is repudiated by the venomous arguments that pepper the literature on hominin adaptations. The second amounts to a conspiracy theory (and paleoanthropologists gossip to a degree comparable to that of the general population). The third implies that even though what we do now is called *evolutionary* morphology, we have actually not yet figured out how to move from morphological data to rigorous tests of what does and does not count as adaptation in the fossil record, nor have we discovered how critical this might be for reconstructing behavior. In the succeeding chapters I will try to make a case for this third premise. I will further argue that if, at least in the short term, we lower our expectations for what we can glean from fossils, we might be better off.

Since books such as this one are not read for suspense, I will foretell a few principal conclusions here and spend the succeeding chapters developing and defending these ideas. First, there are some intractable problems of functional inference from fossil material, such that we will never know some things, no matter how much we might wish otherwise.[2] Second, the reconstruction of behavior and, ultimately, evolution depends on considerations beyond the identification of adaptations (narrowly defined) in the fossil record. Third, despite the appeal and impact of sweeping explanations for human evolution, we will be better served by framing our hypotheses restrictively; i.e., making them easier to refute. Behavioral reconstructions in paleoanthropology are often imaginative. This is both good and bad. It is good because novel hypotheses enable us to recognize and reassess our operative assumptions about evolution and the information content of fossils. It is bad because resources (intellectual and infrastructural) can be wasted chasing speculations that are baseless and have no realistic prospect of being meaningfully tested.

I will argue that the issues raised by the above conclusions are constructively addressed by a consideration of bone development and its interaction with the environment, because this perspective provides a proximate means of identifying the limits of skeletal adaptation and evolution. The ontogeny of bone can tell us what is possible, what is improbable, and which directions of evolutionary change are more likely than others. Application of this knowledge will provide some ground rules for inferring function and behavior that paleoanthropologists talk about but follow inconsistently. I am sure this prescription causes brows to furrow, on the grounds that we have limited fossil data from which to reconstruct skeletal life history. I believe, however, that given neontological data on bone biology, together with the pace at which the "black box" linking genes to morphology is being dismantled, we will soon be in a better position to evaluate competing hypotheses of hominin behavior and adaptation.

In Chapter 2, I further develop these ideas by dealing with some things that most comparative anatomists do not spend much time worrying about: a theory of functional morphology, the empirical basis for that theory, and the epistemological tension between two distinct research paradigms of morphology. If what we do is a scientific enterprise, then it has a theory behind it. Thus it is worth considering whether that theory is sufficiently developed for us to talk about constraint, adaptation, and selection in meaningful ways. This chapter also lays out how a more explicit consideration of skeletal biology—mechanobiology, in particular—can help us understand how bones adapt and evolve.

In Chapter 3, I revisit some of the theoretical issues raised in Chapter 2, but with a narrower focus on specific methodologies that are applied to the hominin fossil record. The central question of whether theoretical concerns and research practice mesh as intended is explored here as well.

In each of the next three chapters, I address functional and adaptive inference in hominin evolution, using contrasting arguments to explore the potential disjunctions between theory and practice introduced in Chapter 3. Chapter 4 deals with the origins of bipedalism, focusing on the question of locomotor behavior in *Australopithecus afarensis*, for which there is a decent fossil inventory. Chapter 5 asks whether there is a singular hominin dietary adaptation that explains the emergence of our subfamily and whether we have the analytical acumen to make that call. Chapter 6 outlines more explicitly my advocacy for a bone's-eye view of hominin paleobiology, a methodological perspective that links bone development to skeletal evolution. A consideration of a highly contested area of research—the timing of the acquisition of spoken language—is used to illustrate the explanatory potential of this perspective. My discussion revisits whether the chin, a feature considered unique to anatomically modern humans, has any adaptive significance or provides any behavioral clues to the emergence of our species. The chapter also presents a fairly simple mechanobiological model of bone growth and development, in order to illustrate the value of a skeletal biology perspective for understanding the challenges of adaptive inference in paleoanthropology. The examples in these chapters are intended to convey what the scope and limits of functional inference are with respect to our fossil record,[3] as well as to serve as a basis for rethinking how we can approach the fossil inventory more productively. I close Chapter 6 with a number of specific recommendations on how we can more concretely align theory with practice.

Finally, in Chapter 7, I revisit Williams's (1966) proposal for teleonomy in a context where we have only the mineralized remains of minute fractions of extinct populations to work from. I argue that this reality severely limits what we can infer from fossil anatomy, but I also note that we are nowhere near the point of knowing where the knowledge ceiling lies. Finding a way to move from biomechanical descriptions to the more nebulous evaluation of inclusive fitness is crucial, even if we discover that building such a bridge is impossible by exclusive reference to the fossil record. What we want to know and what we can know are not identical sets. A reassessment of what we are able to reliably infer can only improve our understanding of the course of human evolution.

Chapter Two

Situating Functional Morphology in Evolutionary Biology

The central postulate of functional morphology is that the form and activity of organisms are beneficial to individuals in some material and tangible way. The discipline's outlook is teleological. For Aristotle, Georges Cuvier, and other practitioners of comparative anatomy before Darwin, this outlook was not problematic and provided reliable guidance for inferring the fit of anatomical traits with the faculties of their possessors. The idea of adaptation via natural selection changed how functional morphology was conceptualized, but it had limited impact on method and practice for well over a century after *On the Origin of Species*.

Cuvier was temperamentally opposed to the need for evolution. For understanding animal form vis-à-vis the "conditions of existence," however, his brand of functional morphology worked perfectly well, because one can link ecology to morphology without recourse to a theory of evolution (E. Russell, 1916). This is possible when the distinction between pattern and process is ignored: adaptive inference is largely an exercise in pattern recognition, and the details of process are more rarely explored. One can thus recruit identical data and link it to either intelligent design or evolution. Morphologists operate under the idea that evolution is the underlying process and base their inferences on the assumption that a good fit between morphology and environment constitutes *prima facie* validation of a historical process of natural selection. The deliberate rebranding of the enterprise as evolutionary morphology (Lauder, 1981) signified an understanding that a correlational criterion of adaptive inference, by itself, is inadequate.

The ubiquity of adaptation is undeniable, yet its identification in specific contexts is problematic. This has been interpreted as a problem of definition and terminology (e.g., Williams, 1966; Gould and Lewontin, 1979; Bock, 1980; Gould and Vrba, 1982). In some cases, the arguments are esoteric and specious. Gould and Lewontin (1979) advocate for the reality of adaptation without natural selection, using the example of sponges and corals "adapting" to local currents during the lifespan, attributing this plasticity as being "purely phenotypic" in origin. If the capacity for plasticity is heritable, then discounting selection is, at the very least, premature. For the most part, however, this literature has encouraged more clarity of thought. Gould and Vrba's (1982) renaming of "preadaptation" as "exaptation" is preferred, by virtue of its rejection of teleology and acknowledgement that preadapted states are fortuitous. Similarly, the concept of umwelt—the details of a species' behavior and sensory capacities (Bock and von Wahlert, 1965)—is typically neglected in comparative anatomical studies but is particularly germane to anthropology. Primates have cognitive tools that, for example, cockroaches do not, which means the former should exhibit greater behavioral plasticity with respect to ecological conditions. This presents an added challenge to the functional morphologist, as tight correlations between anatomy and environment should be harder to come by as a result. Undoubtedly, *Homo* is the worst-case scenario facing the morphologist.

Defining Adaptation: Essential or Esoteric?

The sheer volume of literature on concepts and recognition of adaptation suggests that there is no single definition that will work well across all contexts, and Pigliucci (2003) offers that forcing this objective would be counterproductive. Instead of trying to nail down a hard and fast criterion, we can attempt to circumscribe the concept, so it is operational across different contexts (Reeve and Sherman, 1993). To that end, one can articulate a minimum set of conditions that are necessary, but perhaps not sufficient, for the inference of adaptation. The first condition is that adaptation in the Darwinian sense must arise through a process of natural selection. This is the "historical definition" (Lauder et al., 1993). Williams (1966:109) admits the tautological nature of natural selection in this context but was unworried, because "there can be no sane doubt about the reality of the process." In this view, natural selection is neither law nor principle, but a syllogism. Given variation, superfecundity, and limited resources, the process is mathematically inevitable (Susskind, 2013).

A second criterion is thermodynamic and does not depend on the prior operation of natural selection, although it will certainly apply in most cases. If there is a trait or behavior that does not save energy vis-à-vis alternative states that exist in a population, then its status as adaptation should be met with some degree of skepticism. The idea here is that any savings in the somatic energy budget are available for reproduction, conceived in the broadest sense possible. Objections have been raised. R. Smith (1982:99) rejected the idea as reductionist, because "animals are *different* from machines" (emphasis in original). In other words, natural selection tinkers, and animals are not engineered; at best they are jury-rigged. Yet the logic of natural selection stipulates that relative efficiency is an expected outcome of the process: an above-average management of one's energy budget should have a positive influence on fitness, statistically speaking. Near-optimal solutions need not arise, but energy saved in somatic maintenance has no value in evolutionary accounting unless it is successfully applied to the production of viable offspring. Research on senescence proceeds from this premise (Bribiescas, 2006).

Another objection to a thermodynamic definition of adaptation is that it ignores sexual selection. This critique can be answered in two ways. The first is to argue that the effects of natural and sexual selection are the same, as far as the currency of evolutionary success goes. Traits that can be shown to provide good solutions to ecological problems but do not impact reproduction are not being selected, since "fitness" is always shorthand for net successful offspring in the evolutionary context. An organism that diverts energy to ornamentation at the expense of locomotor performance is not incurring a fitness "cost" at all if, by the time its lack of speed lands it in the jaws of a predator, it has produced more offspring than its conspecifics will over their lifetimes. This constitutes the second answer to the thermodynamic objection: sexual selection is subject to the same energetic accounting as natural selection. The peacock's tail in no way transcends this energetic bind. All that has happened is the peacock has moved some somatic calories to the reproductive side of the ledger. That ledger is the biologist's invention, but the efficient application of energy to reproduction is the only evolutionary strategy that matters. In the context of natural selection, morphological adornments are as adaptive as features that enhance feeding or locomotor efficiency. If adornments endure in a lineage or spread through a population, the alternative explanations are that such a feature has enhanced fitness, arises via pleiotropy,[1] or persists by accident. The latter event can be given credence if there is evidence of small effective population size (easy to assess, in theory) or if direct or indirect selection against the trait is very weak (extraordinarily challenging to demonstrate empirically, perhaps requiring omniscience). The efficacy of drift can be modeled over evolutionary time, and such simulations serve as a useful foil for adaptive hypotheses otherwise immune to meaningful tests (Ackermann and Cheverud, 2004). Adornments that have not plausibly evolved by drift and are shown to have mechanical or performance liabilities should, by default, be given consideration as a signal of fitness in a population. The bearer of the adornment is incurring some nonzero cost, but its observed competence in the environment unambiguously demonstrates that the cost can be borne. The subjective trait value of attractiveness can be reckoned simply in energetic terms.

While traits ordinarily discussed in the context of sexual selection are not defined by biomechanical properties, functional morphological analysis has a clear role to play in understanding their evolution. The supraorbital torus of anthropoids provides an example. M. Russell (1985) applied a beam model to the supraorbital torus to argue that it served as a buttress against masticatory forces in early *Homo*. Though the suitability of the model was questionable,[2] it provided a useful point of focus. Hylander and Johnson (1992) established that peak physiological strains in the anthropoid torus are quite low, which argues against the browridge being an adaptation for structural integrity. Comparative morphometric work (Ravosa, 1988, 1991; Lieberman, 2000) suggested instead that the browridge was linked to the spatial positioning of the viscerocranium relative to the neurocranium. This provides a proximate ontogenetic explanation but does not fully resolve whether the browridge represents a true "spandrel" (Gould and Lewontin, 1979). The possibility that the browridge is an advertisement of fitness has received little attention (Ravosa, 1988; Brunet et al., 2002; Fannin and McGraw, 2018; Fannin et al., 2021), but it is no less satisfactory than alternative hypotheses.

Hylander and Johnson (1992) proposed that the adaptive significance of the browridge is to protect the individual from environmental trauma (e.g., arising in agonistic interactions). Relative to people, nonhuman primates generally do not play well with others, and traumatic injuries, either from conspecifics or falls from trees, are not unknown. The crucial question in terms of adaptation is whether such events—specifically with reference to skull integrity—occur with sufficient frequency for selection to have an impact (an additional example from paleoanthropology is provided in Chapter 6).

The thermodynamic criterion still applies if resisting trauma is the target of selection, but it is less persuasive as the arbiter of adaptation than would be the case for day-to-day activity guiding the selection process. A caveat to the thermodynamic criterion is that efficiency and economy have to be balanced with management of risk. In Hylander and Johnson's (1992) calculus, the cost of the browridge metabolically is low and/or the risk of trauma sufficient to fracture a non-browridged skull is not zero.

Safety factors in bone presumably attest to this need to ensure against accidents. A safety factor is the ratio of a critical stress to the stress experienced in normal activity;[3] the latter is usually referred to as "physiological" stress. Stress here has a precise definition—force applied over a finite area—making it easily comparable across species of different sizes and activity patterns. With respect to vertebrate skeletons, Rubin and Lanyon (1984a) and Biewener (1993) suggest that safety factors may be relatively invariant, with a value between 2 and 4 in many bones.

The presence of safety factors superficially repudiates a thermodynamic criterion, since it is obvious that feeding, locomotor, and postural systems can be designed with greater efficiency and still leave a competent animal. The ubiquity of safety factors can be explained in terms of two possibilities: fatigue or accident insurance. Fatigue is a property of some materials, including bone, in which ordinarily "safe" stresses (and attendant deformations or strains) can produce damage if they are applied repeatedly over time.[4] Thus fatigue is the progressive degradation of bone's structural integrity over a large number of loading cycles, which, considered individually, pose no threat to the skeleton. Fatigue describes many biological and nonbiological materials and can be thought of as a time-dependent closing of the gap between experienced stress and ultimate stress. Fatigue life depends on the interaction of strain magnitude,[5] as well as the number of loading cycles, and experiments on fatigue in bone have exploited this by loading bones to supraphysiological strains over short durations (e.g., Mori and Burr, 1993). While not re-creating a realistic *in vivo* environment, this is a practical solution given that physiological strains could require fatigue experiments lasting months or even years.

Many safety factors in bone are large enough that the fatigue life of skeletal elements exceeds multiples of life expectancy—i.e., some elements appear to be overdesigned for structural integrity, given everyday physiological loads. In such cases, fatigue does not explain a safety factor. If normal use does not threaten mechanical integrity, two possibilities are that (1) an unnamed constraint precludes optimization in terms of material economy; or (2) selection, which cannot possibly preclude all accidents, can nevertheless make most of them survivable. Safety factors represent a functional requirement of structural integrity, but these are presumably determined by a trade-off with an equally important requirement: efficient performance (Currey, 1979). The reality of different safety factors, even within the same bones (Skedros et al., 2003), means that variance in fitness is not uniform across the skeleton (Figure 2.1).

The recognition that adaptation can refer to both a process and a trait presents no special problems that clarity of argument cannot overcome. It is understood that adaptation, as a morphological "thing," requires the process of natural

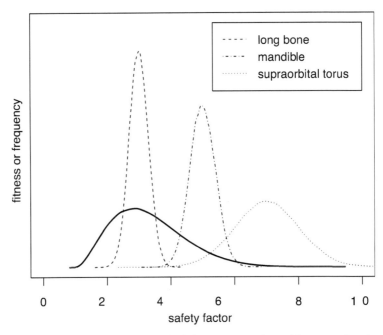

Figure 2.1. Safety factors differ among skeletal elements. Hypothetical frequency distributions are drawn for safety factors of three different bones that experience different levels of peak strain during normal activity. Long bones have the lowest safety factors, because their physiological strains are relatively high. Mandibular strains can be as high as those experienced in long bones but are ordinarily modest by comparison. Supraorbital bone experiences very low strains *in vivo*. A hypothetical fitness function (*solid line*) is provided for long bones. The fitness at a safety factor of 1 is near zero, but the function is not symmetrical, because higher safety factors are merely inefficient, performance wise, whereas lower safety factors have obvious fitness consequences of fracture risk. Presumably, at high safety factors, the performance costs become intolerable, since the mass of such bones precludes needed acceleration across joints. This fitness function is inapplicable to mandibular and supraorbital bone if one expects that the mean trait value is close to a local optimum for fitness (Williams, 1992). For the mandible, the high safety factor might relate to the mitigation of fatigue risk, if the daily loading cycles exceed those of the long bones by a significant margin. For the supraorbital bone, it is difficult to explain the observed safety factors in terms of fatigue risk; Hylander and Johnson (1992) instead suggested that the high safety factor in browridges is attributable to selection against traumatic events. This is reasonable if the frequency of such events is sufficient for selection to act.

selection, provided such a thing is heritable. If a trait is good in some subjective way or confers boundless fecundity but did not arise via natural selection, it is not an adaptation by the historical definition, no matter how beneficial it is in the present moment. It is easier to detect the benefit of a trait than it is to demonstrate that selection produced it, but this is a methodological problem, rather than a theoretical incongruity of the process and its outcome. The collective insistence on more formal criteria for recognition of adaptation has created some communicative difficulties. Understanding the current utility of a trait or behavior may or may not identify it as an adaptation, and one's ability to link the trait to better fitness is not decisive, either. The selective force that brought the trait into existence is rarely known, and in any given lineage or clade, an adaptation may have limited durability in terms of its original ecological application. This should surprise no one contemplating the vagaries of environment and evolution, but it presents problems for answering the question of original selective context. If environmental conditions change, such that morphology is recruited for new purposes, adaptations that persist will usually qualify as exaptations.

These problems are largely operational, rather than stemming from any incoherence of theory. The relationship of selection to adaptation and adaptation to fitness is the same under divergent perspectives of, for example, gradualism or punctuated equilibria. The practical issue is that adaptation is inferred, based on interspecific differences among taxa, but diagnosing adaptations proceeds from the logic that variance in fitness within populations and species is what gives rise to them. In the calculus of fitness, what mattered to *Paranthropus* individuals was their behavioral strategies vis-à-vis their congeners; if *Homo* was sympatric and was exerting ecological stress on the australopiths, what was still decisive in their evolution was their response to the stressor, relative to others in their population. This raises the question of whether species-level adaptations are a necessary or desirable metric for functional inference. It would appear that fitness must be defined as a species-level property in order for the idea of species-specific adaptations to be valid in an evolutionary context, but this ignores population variance in traits, an essential requirement for adaptive evolution. This issue is further discussed in Chapter 3.

Traits can persist and be employed for individual benefit, even if the original stimulus for selection has long since disappeared. From this observation, adaptations (historically defined) compose only part of a species' behavioral repertoire. The concept of biological roles fills the gap. Bock and van Wohlert's (1965) conception envisions the biological role as interacting with a selection force ("synerg") to produce adaptation, with the sum of the synergs defining the niche. Biological role was not synonymous with function in Bock's (1980) scheme, though the distinction is unimportant in many contexts. Radinsky (1985) views a biological role as how a feature or complex is utilized by an organism. It has no necessary status as an adaptation, nor can it be diagnosed by an inference of functional performance in the abstract. Adaptations are a subset of biological roles at any given time. Bock (1980) conceptualized this in more processual terms, seeing biological roles as adaptations in the process of either disappearing or coming into being.

The biological role of a trait or behavior is not merely a means to label things that are not adaptations. Biological roles define behavior and performance in a broader context, which is possible under the maligned adaptationist paradigm. If one's goal in paleoanthropology is to understand hominin behavior, the study of historical adaptations, strictly defined, is inadequate. Biological roles encompass the concepts of functional (= physiological) adaptation and phenotypic plasticity, neither of which fits neatly into the classical concept of adaptation (Ravosa et al., 2016). The study of early hominin behavior is more productively focused on biological roles rather than on adaptations. Confining inquiry to the identification of adaptation, strictly in terms of its historical definition, is arguably too narrow to understand hominin evolution. This theme is further developed in Chapters 4 and 5.

The aspiration to focus on biological roles, including those that qualify as adaptations, creates new challenges. Under this perspective, traits are no longer applicable only for a particular function. Instead, they are capable of tasks other than those filtered by natural selection. While this realization is helpful in theory, if selection has a less direct influence on biological roles, their inference from traits themselves would seem to be more difficult. But identification of selective forces in functional morphology is rare, so the added difficulty may not be particularly dire.

Laboratory data on mechanical performance have proven invaluable for modeling the function of anatomical systems, but this information does not ordinarily encompass behavioral activity in natural settings. What looks certain to be adaptive significance in the laboratory may turn out to be a trivial capacity in the field. Performance in the lab alerts the investigator to a particular capacity, but it may not be a critical aspect of ecological behavior. Primates—hominids, in particular—are maddening creatures to study in this context. While they must obey the laws of physics, their behavioral plasticity is such that they can solve ecological problems through means other than physical exertion. An observation of superior performance is likely to be indicative of an adaptation of some sort, but it is not necessarily the one the researcher has in mind. The buttressing effect of a supraorbital torus can be quantified mechanically, such that the structural benefit is obvious, but this does not have any bearing on its adaptive significance. The biological role may be different, even if the mechanical analysis is correct in all its details, or a supraorbital torus might simply be an obligate but inefficient osteogenic response to mechanical stresses arising elsewhere (Menegaz et al., 2010).

Repackaging the Enterprise

The scope of functional morphology, in Alexander's (1987) view, is to describe how anatomies work, under the assumption that observed designs were brought about by natural selection. Alexander usefully differentiates functional from ecological morphology by the nature of the comparisons being made: the performance capacity of the shoulder joint of *Homo habilis*, in terms of stability versus mobility, is the domain of the functional morphologist, while a comparison of glenohumeral joints among early *Homo*, with a focus on differences in habitat use and local environments, qualifies as ecological morphology. Evaluating performance by the application of engineering or physics is central to both endeavors. Most of the time "good" performance is considered indicative of "good" fitness, but Lauder et al. (1993) argues that this is not logically necessary, given examples of high-performance traits with net deleterious pleiotropic effects. Ideally, evolutionary morphology is ecological morphology, with the imposition of phylogeny as a check on unwarranted adaptive speculation. Lauder (1981, 1982) views phylogeny as the window into intrinsic constraints, which are the lacunae that plague functional morphology as typically practiced. Many contributions to the primary literature invoke the phrase "evolutionary morphology" to describe work that is purely functional in scope; any mechanically descriptive study with post hoc behavioral inference would qualify. With the advent of purely molecular systematics, however, evolutionary morphology can proceed from a quasi-independent assessment of phylogeny and morphology. The difficulties of assessing adaptation in paleobiology remain, but evolutionary morphology would appear to have a more explicit explanatory foundation.

The foregoing raises the question of whether functional morphology, as a theoretical perspective, is sterile with respect to biology. Perhaps it is a matter of defining premises precisely. What evolution and ecology entail is apparent, even if there are disagreements about finer points of domain and content. Definitions of function are rare and, when given, as often as not are unhelpful. S. Wainwright (1988) sees it as the change in structure with respect to time. Bock (1980:219) views it less concisely as "all physical and chemical properties of a feature arising from its form including all properties from increased levels of organization." Both are unobjectionable and, arguably, correct, but what these definitions facilitate in practice or understanding of organismal biology is not obvious. The question, How does it work? is instead fairly straightforward. The principal problem with function in the evolutionary context is that it is often unintentionally articulated in teleological terms. This was fine for Aristotle and Cuvier, but it should not be acceptable for those committed to methodological or philosophical materialism. Ruse (2000) demurs, believing that we appreciate the metaphorical value of teleology for describing adaptation without endorsing any metaphysical undercurrents. O'Grady (1984) suggests that function in the biological realm be described as teleonomic (end-seeking activity), but that evolutionary processes be described teleomatically (reaching an end through physically inevitable events). These arguments do not mean we have a poor understanding of what function is. On the contrary, they implore us to be mindful that in its invocation, there is an important distinction between efficient and final causes.

Most of us practicing functional morphology would chafe at the judgment that it is merely descriptive, but this is the enterprise in its purest form. It is worth repeating here that functional morphology does not require evolutionary theory. An evolutionary explanation, however, necessitates more than the establishment of correlations between anatomy and ecology: phylogenetic, ecological, and developmental data must also be recruited. This has been understood for decades, but paleoanthropology—in practice, but not in theory—has been slow to incorporate these data. This is explored in Chapters 3 through 5.

Form versus Function: The Question of Primacy

The "functional" qualifier to morphology is not specious; S. Wainwright (1988:678) observed that "no one studies just morphology." In the first half of the 19th century, a vigorous debate on the proper emphasis in comparative morphology dominated organismal biology. Cuvier was the standard-bearer of the functionalist school of thought; its principal aim was establishing the fit between anatomy and the "conditions of existence" (McBirney and Cook, 2009). The structuralist school (represented by Étienne Geoffroy, Richard Owen, and others) regarded the discovery of homology as paramount. From an evolutionary perspective, the transcendentalist's obsession with archetypes is difficult to comprehend, while Cuvier's functionalism is more palatable, as long as ultimate causation is ignored. Even so, structuralism is alive and well for two (not altogether distinct) reasons: (1) its essentially correct concept of homology is critical for the inference of both phylogeny and evolution, and (2) its theoretical foundation is superior to that of functionalism. This second point is simply due to the recognition of constraint as a potent foil for adaptive evolution.

Dwyer (1984) contrasts the two approaches to morphology as having distinct axioms. Functionalism holds that "all forms are possible," while structuralism is premised on the idea that "all environments are possible." The former generalization is instructive, while the latter is unnecessary to understand the appeal of structuralism and, in any case, is impotent as a heuristic. Under a functionalist paradigm, constraint has no meaningful role to play in the generation of form: natural selection and drift are its primary determinants. Optimization of morphology is possible—in theory—through the relentless and unfettered operation of natural selection. If optimal form is not observed, it means that environmental fluctuations have occurred too rapidly for selection to keep pace. Drift can produce "accidental" morphology, and, as a result, the functionalist approach cannot be fairly labeled pan-adaptationist in its outlook. Specific traits, however, are rarely judged as products of drift. As Gould and Lewontin (1979) surmise, when traits do not fit with inferred ecological requirements, the standard recourse is to reassess the nature of the ecological pressure. Functionalism has endured in large part because it has been tremendously successful for discovering adaptations; the most critical question is whether these discoveries, though often intuitive, are consistently correct. While this remains uncertain, if functionalism is valid, then natural selection should suffice for explaining morphological variation in most contexts. This would mean that, but for the odd exception, no other evolutionary forces are effective in producing morphological change.

Structuralism, perhaps ironically, sees the organism as an active agent vis-à-vis the environment. Individuals, to some degree, select their environments; if they choose poorly, there will be no such decision to be made by their progeny. In the functionalist domain, the environment shapes the organism. Evolution will be ecologically driven, such that if the environment is stable, at some point no meaningful morphological change will occur. This distinction can be clarified by example.

For the sake of argument, let us assume the pejorative "bipedal cow" hypothesis for *Paranthropus boisei* is correct. This idea arose from isotopic studies, which revealed that East African *Paranthropus* teeth had a carbon isotopic signature, indicating consumption of C4 resources, which could be plant matter or animals that consumed such plants (Cerling et al., 2011). C4 plants include grasses and sedges, rather strange food choices for primates (with the exception of *Theropithecus*). As most paleoanthropologists doubt that australopiths were committed carnivores (although see Cachel, 1975; Szalay, 1975), the isotopic data suggest bulk consumption of grasses and sedges in *P. boisei*. While the teeth of these hominins appear to be poorly suited for shredding grass in comparison with specialized grazers, the hypertrophied masticatory apparatus for *P. boisei* does not preclude any diet in the primate repertoire. There is no food imaginable that could not be processed, only things that could be masticated efficiently or inefficiently. By several accounts, the East African environment was not spatially or temporally homogeneous (Kappelman, 1984; Feibel et al., 1991; Reed 1997), and there is no reason to doubt that *Paranthropus* could move among habitats.

The fact that facial morphology in *Paranthropus* is highly derived has led to the reflexive conclusion that the genus represents an ecological specialist by hominin standards (Wood and Strait, 2004). As Ungar et al. (2008) note, Liem's paradox applies here. The paradox is the observation that highly derived morphology, which presumably was selected for some aspect of ecological specialization, is sometimes encountered in animals which may exhibit very generalized behavior, in terms of foraging strategies or dietary breadth. This bias precludes recognition of fossil species that could potentially exploit multiple microenvironments over short and long time frames. *Paranthropus* had the capacity to choose its environment within the constraints of its mobility, and its success can be seen as an indicator that for over a million years, it was not making terrible choices. This encapsulates the structuralist perspective of Dwyer (1984): an organism has limited means to evolve quickly; therefore, survival depends on seeking out a workable habitat. Constraint is assumed to be a real and ubiquitous problem. The functionalist perspective is more straightforward: if *Paranthropus* is consuming grasses all the time, then that is what it is adapted to eat. Natural selection will solve a problem, given sufficient time and a suitably variable gene pool. Paleoanthropologists understand that if this is axiomatic, then adaptive inference for fossils is a matter of paleoecology. The fit of anatomy to ecological variables is then mandatory. In this view, the *Paranthropus boisei* gnathodental complex is an efficient solution for processing C_4 plants. The problem, of course, is that better grass-processing machinery can be imagined and observed (the wildebeest grazing nearby gives one pause).

If the hard-and-fast commitments of the purist are relaxed, there remains a difference in perspective that is still useful. The explanation from the structuralist viewpoint is that *P. boisei* is eating grass because (1) it is available, and (2) it had the means to extract energy from that resource. The behavior can be conceived of as an evolved adaptation, while the teeth and jaws doing the work are more properly considered exaptations. The functionalist sees the dietary challenges as requiring an adaptive—i.e., driven by selection—solution. In this case, invoking exaptation as the explanation of morphology is still sensible, and there begins the slide into a structuralist worldview.

Functionalism, in its unadulterated form, is incompatible with evolutionary theory. Structuralism is on a more secure footing, because constraint is acknowledged and observation refutes the central prediction of functionalism: if environments vary more or less continuously, then evolution should produce a continuum of form. Morphospace is discontinuous, however, which leads to the rational conclusion that morphological solutions to ecological problems are finite (Raup and Michelson, 1965; Raup, 1966; Alberch, 1980; Thomas and Reif, 1993). This suggests that any defense of functionalism is merely a reactionary argument of no consequence to a science of morphology. A functionalist perspective is utilitarian to a fault, but in many contexts it will successfully identify adaptations. There is probably a very good reason why biological anthropologists have not embraced the hypothesis that forelimb morphology in *Hylobates* was selected for some other reason than ricochetal brachiation. Functionalism is inert only if, in general, natural selection cannot overcome the noise generated by non-Darwinian factors and override apparent constraints. This is a tall order. Functionalism under a different name (the "adaptationist paradigm") was recruited for repeated scorn by Gould (1980; Gould and Lewontin, 1979; Gould and Vrba, 1982), although he conceded the necessity and power of natural selection to complete his punctuated equilibria model of evolutionary change (Gould and Eldredge, 1977). Since, in comparison with the structuralist perspective, functionalism is theoretically bereft, it is something of a theoretical headache, since it seems to work anyway. Paleoanthropologists are well versed in structuralist principles (even if they do not identify them as such), but succeeding chapters will show that in practice, functionalism remains influential.

The Formalization of Structuralism

The rise of constructional morphology as a theoretical paradigm for morphology incorporated the tension between structuralist and functionalist perspectives into a more complete framework for a science of biological form. The basic idea, originally floated by Weber (1954) and more fully articulated by Seilacher (1970), is that there are three aspects of form-generation in evolution: the functional-adaptive, the historical-phylogenetic, and the morphogenetic-fabricational (K. Vogel, 1991). Arguably, the historical and morphogenetic aspects imbue structuralism with tangible processes from which the concept of constraint becomes less abstract and unwieldy. The need for understanding ontogeny is explicit, and the practice of correlating adult forms exclusively to environmental variables is exposed as incomplete.

Constructional morphology has had relatively little practical influence on paleoanthropology. There are, minimally, two reasons for this. The first is that the operationalization of the concept involves what are—for students of hominin evolution—unfamiliar examples from invertebrates (e.g., Seilacher, 1991), and the second is that reliable tests in a paleobiological context would be hard to come by (Lauder, 1981). Constructional morphology very clearly anticipated Gould and Lewontin's (1979) critique of pan-adaptationism. The constructional morphology thesis is that natural selection has a limited influence on form, and the scope of this influence is discoverable. Structure is seen to have both proximate and ultimate material causes,[6] and it evolves through modification of development.

As with any vibrant model, constructional morphology has been modified from its original conception. Seilacher (1991) added a factor of "effective environment" to the trinity of agents and thus transformed the 2-D triangle of form-generation into a tetrahedral pyramid of biomorphodynamics. Raup (1972) eschews the geometric analogy altogether and identifies five principal factors "controlling" morphology: (1) historical/phylogenetic, (2) functional, (3) ecophenotypic, (4) structural, and (5) chance. The ecophenotypic factor encompasses what are discussed today as reaction norms and phenotypic plasticity. Raup's structural factor subsumes both constraint and its counterpart, "opportunities" (in the sense of S. Vogel, 1988). The addition of chance to the mix suggests that there is a residual variance one can quantify and, in so doing, specify the degree of stochastic influences. For morphological analysis, there are statistical techniques that partition variance due to phylogeny from other sources. These methods summarize the patterning of traits in populations but do not directly reveal anything about biological roles. Thus variance components attributable to function are not easily identified, other than as a residual value of some kind. Operationalizing the several versions of constructional morphology is challenging, and while the theory behind it suggests that quantifying the myriad influences on form is possible, convincing demonstrations are elusive. Raup (1972) employed simulation modeling—the basic form of inquiry in theoretical morphology—to demonstrate its promise as a productive avenue of investigation; this has been prescient with respect to paleoanthropology (see Chapters 3 and 6).

On the face of it, constructional morphology may appear to be a more holistic approach to important questions in morphology, but holism in comparative anatomy generally represents a reaction to atomism—the isolated analysis of traits without much regard for the integrated organism (Dullemeijer, 1980). Application of constructional morphological principles would seem to be as challenged by a holistic aspiration as by more-conventional functional morphological sensibilities. K. Vogel (1991) argues that biomechanics underlay all three aspects of constructional morphology. Ontogeny is reduced to force and energy; function (whether as adaptations or biological roles) is transparently mechanical. The phylogenetic aspect is built through "a series of mechanically coherent constructions" (K. Vogel 1991:64) that appears to be identical to the mechanical underpinnings of ontogeny. At the same time, he argues for an integrative approach; i.e., a bias toward holism, because atomism (the very physical reductionism he advanced) must entail a loss of information about the individual, which, in morphological circles, is the implicit vehicle of selection. This issue is more nebulous than is generally appreciated. Dullemeijer's (1980) appeals for a more holistic outlook never managed to present a comprehensive or accessible accounting of function for a whole organism. Mayr (1983) considers holism to be "nonexplanatory" with respect to adaptation.

Advocates of structuralism argue that it resolves the conflict of atomism and holism (see Webster and Goodwin, 1982; Reif et al., 1985), but there is no holistic benchmark to demonstrate the inadequacy of an atomistic analysis. Perhaps this is just an operational problem, soon to be resolved by adequate simulation and corroborated by experimentation, but it is presently hard to tell what that holistic analysis would look like. More importantly, what would we learn? There is undoubtedly some connection between the mechanical minutiae of the transition of the cardinal to the caval venous system, on the one hand, and the ossification of sclerotomes, on the other, but it is difficult to discern how its discovery would cause us to rethink the tissues' adaptive utility. If the answer is framed in terms of thermodynamics—energy devoted to one process is unavailable for another—then we are merely reiterating Aristotle's concept of "excess and defect," only from a different perspective. This underscores the idea that a thermodynamic criterion is necessary but insufficient for inferring adaptation.

Structuralist sensibilities also productively focus on the developmental context as the means by which to comprehend morphological innovation (Rieppell, 1990; Müller and Wagner, 1991; D. Peters, 1991; Plotnick and Baumiller, 2000). It is easy—and wrong—to explain the appearance of a novel trait as the product of an extrinsic environmental stimulus. That stimulus cannot logically select for the emergence of something that is not yet present; it can only maintain or augment it once it appears. Functionalism, in its pure form, cannot explain novelty at all. The morphogenetic perspective that structuralism demands also informs questions of the independence of traits through examination of developmental integration. This approach has been successfully applied to the hominin fossil record (e.g., McCollum, 1999).

Paleobiology and Uniformitarian Principles

The foregoing concerns are exacerbated in paleobiology, because there are data that are inaccessible. Physiology and embryology are irretrievable, except through invocation of sets of assumptions, many of which are justified solely by phylogenetic parsimony. In paleoanthropology, parsimony is a principle of last resort for deciding among alternative hypotheses. As Truzzi (2007) points out, however, there is no rational basis for assuming that the universe operates via parsimony, and it is incompatible with neo-Darwinian evolutionary theory (Dullemeijer, 1980). Its validity, as such, springs from the fact that we agree to use it and can expect it to provide correct answers at least part of the time.

A uniformitarian assumption is necessary for functional inference in paleontology. There are two reasons—both misguided—to deny this. The first is to assert that paleontological data are richer than neontological data and that, as a consequence, we can operate under a relaxed set of methodological restrictions. While it is true that the temporal dimension of paleontological data eclipses those of neontology by orders of magnitude, the inference of function from anatomy is still more reliably drawn in neontological settings, for the simple reason that behavioral variation is observable. The second reason for rejection of uniformitarianism is that its assumption of coherence of process through time is wrong in a particularist sense. As a prescription for practice this assertion is indefensible because it confuses historical events with the myriad but omnipresent forces that produced them. The uniformitarian assumption does not posit that landscape, biotic diversity, or demographic patterns were similar to those existing today. It only maintains that the processes producing these patterns operate via the same mechanisms, irrespective of time.

An insistence on uniformitarianism restrains our imagination productively, such that there is consensus that a uniformitarian assumption is a prerequisite for an intelligible science of paleobiology. Paleoenvironmental models are completely dependent on it, and these reconstructions, analyzed in conjunction with fossil morphology, determine paleoecology. Associations of specific morphology with particular environments today are used as analogs for functional inference in paleontology. This can present problems if there are limited modern analogs available, because a fixed association between paleoecology and morphology follows (i.e., the inference is nothing more than observation of correlation). Such an association may or may not be obligatory. Addressing this issue in the paleoanthropological record is problematic, because the sparseness of data prevents anything but the coarsest reconstructions of ontogeny and morphological variation, in addition to uncertainties regarding environmental variation. This means there are some questions that are intractable.

Understanding morphology in paleoanthropology is dependent on a firm grounding in dental and skeletal biology; other aspects of physiology beyond those explicable on allometric principles are largely invisible.[7] Fortunately, bones and teeth are capable of yielding a wealth of data on life history. They have also been studied thoroughly to establish functional linkages between locomotion and skeletal anatomy, as well as diet and dental morphology (Fleagle, 2013; Gebo, 2014). There is no reason to believe that such associations have not existed throughout primate evolution, but the critical question is whether the modern primate data set—which has its own idiosyncrasies in morphological diversity—is completely suitable for comparison in time-averaged, taphonomically biased paleontological contexts.

A uniformitarian assumption requires evolutionary processes to work similarly at any time and place, and it holds that local biotic and environmental conditions impact morphology over the short and long term. Since these local conditions are incompletely known, outcomes are almost always unpredictable. An important point of disagreement about how the evolutionary process works is the relative efficacy of natural selection in creating morphological diversity. Phyletic gradualism views natural selection as temporally ubiquitous, but the unidirectional nature of selection is what makes the model an imagined paradigm. From the perspective of morphology, it is arguable that patterns of diversity would be distinct under the competing models of phyletic gradualism and punctuated equilibria. Natural selection will be ubiquitous in both, provided that ample genetic and phenotypic variation exists. The principal difference is that under punctuated equilibria, natural selection is particularly powerful in producing change during instances of speciation. It is not absent over a species' duration, but its role will be stabilizing, for the most part. The structure of the punctuated equilibria model is such that morphological change and speciation are coincident. This is most convenient for the paleontologist, as it resolves taxonomic, adaptive, and phylogenetic questions simultaneously. Levinton and Simon (1980) and Levinton (2001) provide detailed arguments for skepticism about the temporal conjunction of speciation and adaptation.

One other conceit of the punctuated equilibria model is that the fossil record can be taken at face value. This, in essence, is the idea that gaps in the stratigraphic record do not conceal true rates of morphological change, and that "stasis is data" (Gould and Eldredge, 1977). Haldane's paradox offers a different perspective.[8] This paradox is that the rate of evolution in the laboratory is several orders of magnitude higher than that observed in the fossil record. It is explained as the result of sampling error, inherent in paleontology (Figure 2.2). If most adaptive events are, in fact, invisible in the fossil record, given a null hypothesis of stochastic environmental variation, connecting the dots of an evolving trait will yield an estimate of evolutionary rate that misses most of the evolution—specifically via natural

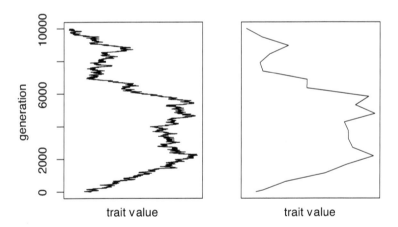

Figure 2.2. Haldane's paradox is illustrated as the evolution of a trait mean in a hypothetical population. The *left panel* is a highly improbable, but desirable, paleontological record in which every generation (synonymous with time) of the population is sampled. Evolutionary rate is the shift in the trait mean value across generations. The *right panel* is a sampling from every 500th generation of the same hypothetical data set. Between generations 2,500 and 4,000, and again between generations 6,000 and 6,500, evolutionary rate estimates will be different between the two sampling regimens. Poor temporal sampling—a property intrinsic to the hominin fossil record—will almost certainly underestimate evolutionary rates, unless selection is rapid but consistently directional. Note that apparent stasis need not reflect an absence of either stabilizing or short-term directional selection. Drift may be operable as well, but diagnosing it versus selection is problematic by exclusive reference to temporal changes in the trait mean.

selection—that has actually occurred. If adaptation by natural selection is a potent agent of this evolutionary change, this means that most records of adaptive events will be inaccessible paleontologically. This does not necessarily compromise recognition of adaptation, given the tests that can be applied (see Chapter 3), but understanding rates of adaptive evolution for specific traits will rarely have adequate empirical support. Better temporal resolution (left panel of Figure 2.2) mitigates the problem of discerning between episodic versus monotonic rates of adaptive evolution, but even in the case of a "complete" fossil record, we would be severely limited in our ability to identify selective agents from the paleontological context.

Allometry as Explanation

Allometric analysis is both a method and a theoretical perspective: it is descriptive and explanatory. In the best cases, an allometric study streamlines interpretation, so morphological trends that are physiologically obligate can be distinguished from those that relate to the specifics of a species' ecology. If body size is indeed the most important ecological variable in primate biology, then there are many opportunities for useful, if general, inferences in the paleontological context. In primates alone, body size has predictable effects on diet (Kay, 1975; C. Ross, 1992; Ungar, 1998), locomotion (Jungers, 1984; Fleagle, 1985), reproduction (C. Ross, 1988), and other aspects of life history (Harvey and Clutton-Brock, 1985). Scaling of the primate skeleton and dentition has been exhaustively explored. In comparative multivariate studies of morphology, size is usually the default explanation for patterning taxa along the first principal component. Cardini et al. (2019), however, warn that apparent separations of taxa (or other operational groupings, such as ecotypes) may be spurious in cases in which high dimensionality of the data (i.e., lots of variables) is analyzed over comparatively few groups. In geometric morphometric analyses in paleoanthropology, this combination is likely (Figure 2.3). More importantly for investigations of function and adaptation, Cardini et al. (2019:303) caution that in interpretations of multivariate ordination, "there is no reason for the axes themselves to be especially meaningful biologically."

Absolute size has inarguable biological consequences, but what garners the most scrutiny in the literature is assessment of relative size, with undercurrents of adaptation. It is both true and uninteresting to point out than an orangutan femur experiences higher load at failure than that of a tarsier. The ratio of bone strength (stress at failure) to bone stress during habitual activities is far more important for understanding locomotor behavior. Similarly, there is nothing remarkable about the size of *Paranthropus boisei* molars by a mammalian criterion;

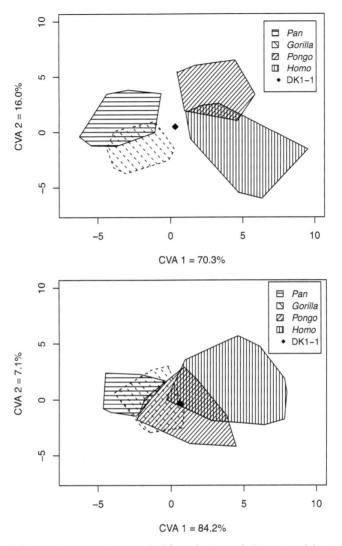

Figure 2.3. Ordination is a convenient method for reducing multidimensional data into accessible form. It is also used to inform functional inferences, based on the location of fossil specimens in multivariate morphospace. Whether this patterning in morphospace is delineating actual performance differences is unclear, since the multivariate distillation of individuals does not provide biomechanical information specific to particular behaviors. In this case, a scapula of DIK-1-1 is unique when plotted on a combination of linear and angular measurements (*top*), but it is situated within the ranges of three extant taxa when using exclusively angular variables (*bottom*). The ordination is depicting the fossil form accurately, but the adaptive significance is not transparent, beyond taxonomic proximity. This may underlie Oxnard's (1978) curious reference to these approaches as "hypothesis-free methods"; i.e., the patterning of form based on a multitude of measurements may not be ideal for addressing specific biomechanical questions. *Note*: The species polygons are for adult and juvenile specimens of extant large-bodied hominoids, and samples are not identical between plots. Adapted from the canonical variates analysis of D. Green and Alemseged (2012).

what needs explaining are their dimensions relative to this hominin's incisors, its facial skeleton, and its body size.

An important question with respect to comparative primate morphology is the role that allometric analyses have for adaptive explanations. Slopes and intercepts summarize the numbers at hand in an accessible way; they reveal patterns of variation that can be linked to some aspect of scale. The issue is whether this pattern necessarily provides unique insight into identifying adaptation (Jungers, 1984). There is an unstated but unjustified optimism that this is the case (R. Smith and Jungers, 1997).

Nevertheless, allometry has proven to be an effective lens for understanding skeletal biomechanics. A cogent example is limb bone scaling in mammals (Alexander et al., 1979; Biewener, 1982; Rubin and Lanyon, 1984a). Mammalian limbs scale approximately geometrically with body mass; i.e., the diameter:length ratio is conserved. Unless the stiffness of bone material compensates in larger animals—which appears to be unlikely[9]—this state of affairs appears nonsensical at first glance, because under the assumption that locomotor loads scale in direct proportion to body mass, small mammals are running around with needlessly massive long bones and/or large mammals should be breaking these bones all the time. In purely geometric terms, bone thickness scales too slowly with body size, such that locomotor stresses are ever increasing in larger species. The observation that large mammals have adequately functioning bones indicates that the assumption of a one-to-one scaling of mass to loads is incorrect. The reason gorillas do not have compromised skeletons is that they do not leap around like marmosets. Conservation of bone shape is observed, but this description, in allometric terms, does not translate seamlessly to the discovery of adaptation. What must be explained is why large terrestrial mammals have competent skeletons. The scaling of bone geometry alerts us to the problem (i.e., the particulars of scaling are not the adaptation); the solution (or adaptation) is that there are behavioral and physiological adjustments in locomotor dynamics.

The preceding example underscores that there are physical and geometric imperatives that are functions of scale. This can be described alternatively as a "criterion of subtraction" or "functional equivalence" across the reference variable for size. The criterion of subtraction implies that allometry somehow allows size to be extracted from an analysis, and what remains is "shape." In this case, it is important to recognize that there is no necessary congruence of statistical size removal (or, more properly, correction) with the more nebulous idea of partitioning size and shape biologically. What functional equivalence means is always context dependent. When equivalence is explicit, the concept is useful. Maintenance of skeletal stress levels across various body sizes would be one example.

Scatter about regression lines, despite their status as statistical noise,[10] represents a secondary signal of great informative potential. Larger scatter rightly prompts skepticism that there is any primary signal at all, or, at least, indicates that there are other variables to consider. A taxon's status as residual invites speculation with respect to species-specific adaptations, but there is no blanket justification for such inferences (R. Smith, 1980). The statistics involved require each datum to occupy some point that is not on the regression line. By insisting that a best-fit line represents a state of functional equivalence, one can argue that a species' position in bivariate space is indicative of an adaptation by a specific appeal to relative performance. But this demonstration is insufficient for adaptive inference without some appreciation of behavioral or ecological context. C. F. Ross et al. (2002) provide a thorough deconstruction of the misplaced logic of functional equivalence in allometric analysis.

As has been widely appreciated, one means to ensure robust r^2 values (reflecting the variance in the response variable explained by variation in the predictor variable) is to have an expansive data set, encompassing body-size variation over several orders of magnitude. In such cases, the slope of the line in logarithmic space can be related to basic physical or physiological principles, such as the maintenance of surface area:volume ratio or metabolic rate. Because the best opportunity to discover meaningful effects of scale in an allometric analysis is to compare taxa that share a similar bauplan, phylogenetically restricted analyses are preferred. One feature of the hominin lineage is that body-size variation is relatively modest, probably within an order of magnitude (Grabowski et al., 2015). This means that for hominin (or hominid) studies, the most suitable approach is one of narrow allometry (R. Smith, 1980).[11] This will ordinarily deny the investigator the luxury of small confidence intervals with which to reject hypotheses of isometry or allometry, but this provides some justification for the often-encountered comparative samples with which fossil assemblages are evaluated: modern humans, chimpanzees, and perhaps other Great Apes. More-expansive samples can be applied, but the illusion of improved statistical certainty is offset by a removal of phylogenetic and ecological framing.

Figures 2.4–2.6 show examples of what can be gleaned from a relatively narrow allometric approach and also provide some appreciation for interpretive pitfalls. In each figure, the response variable is relative bone strength, calculated from compact bone geometry of the postcanine mandible. Three reference variables for size are considered, and they represent different points of emphasis. The goal is to ascertain whether *Australopithecus sediba* and *Homo floresiensis* are functionally equivalent to either modern *Homo* or australopiths,[12] using allometry to factor in the effects of scale.

When bone strength is described in terms of bone area (Figure 2.4), the functional question is one of the economical and efficient use of material. All fossils fall in with the modern comparative sample of Great Apes and humans, and there is no obvious transposition of any taxon above the remainder of the sample. Visual inspection suggests that a single scaling relationship describes the sample, so

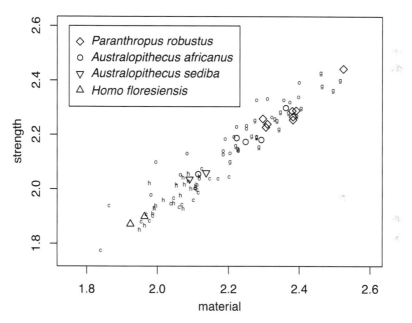

Figure 2.4. The relationship of bone material utilization (cortical bone area) to jaw strength (polar moment of inertia) in the postcanine mandibular corpus of living large-bodied hominoids, South African australopiths, and *Homo floresiensis*. Extant hominoids are indicated by lowercase letters: h = humans, c = common chimpanzees, o = orangutans, g = gorillas. The relationship is shown in $\log_{10}$ space, with the polar moment raised to the 0.5 power to achieve dimensional similarity. This relationship describes how economically bone is deployed in the mandible to achieve load resistance. There is nothing unusual about the fossil taxa relative to living forms, and the relationship is isometric (slope not different from 1.0). The polar moment of inertia depends on both the amount of bone and its distribution about a section's center of gravity. Technically, the polar moment of inertia measures structural stiffness, and it is used here as a general all-purpose variable of load-resisting capacity. Engineers will object to my use of the term "strength" in the y-axis title, but high polar moments of inertia translate to a higher breaking stress (a stronger jaw), all else being equal. Data from Daegling et al. (2014, 2016).

the question becomes whether this also exemplifies functional equivalence. The evidence for this would be a slope of isometry (i.e., 1.0, given dimensionally equivalent variables), which would imply that, on average, large hominins deploy the same amount of bone to achieve a given strength. The empirical slope in this case is 0.98, which is very close to isometry. Consequently, an interpretation of functional equivalence with respect to the efficient use of tissue is reasonable, but this interpretation should go no further.

With Figure 2.5, the same strengths are regressed on mandibular length, which is regarded as the appropriate proxy for evaluating structural integrity of the jaw (Hylander, 1985; Ravosa 2000). The justification for this proxy is that jaw length will be proportional to the bending moment arm acting during feeding behavior. To maintain an adequate safety factor, strength should increase in longer mandibles. The uniqueness of South African australopiths is obvious in this comparison, as their mandibles are generally stronger than living apes at a given jaw size.

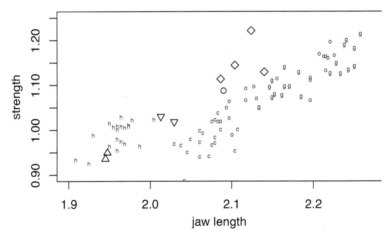

Figure 2.5. The polar moment of inertia (expressed to the 0.25 power), regressed on jaw length for australopiths, *Homo floresiensis*, and modern large-bodied hominoids. Symbols, as in Figure 2.4, are plotted in $\log_{10}$ space. Jaw length is a surrogate for the bending moments impacting the mandible. This variable addresses the "all else being equal" caveat of Figure 2.4. Longer jaws must have higher polar moments of inertia to maintain equivalent stresses. In this plot, it is clear that the fossil slope differs from the living hominoid slope. What is more immediately relevant to functional interpretation is where the fossils lie relative to living forms of the same size (i.e., nearby *x*-axis values). The australopiths have much stronger jaws than similar-sized living forms, meaning their mandibles could tolerate higher loads when biting and chewing. *H. floresiensis* is situated at the lower end of the human distribution. Though functionally similar to modern humans by this criterion, there is no confusing the Flores jaws with ours.

H. floresiensis is situated within the modern human range. *Au. sediba* jaws are strong relative to similarly sized chimpanzee mandibles but weak compared with a modern human standard by virtue of their greater length. Their location in bivariate space does not yield immediate insight into feeding biomechanics, and other lines of evidence present contrasting inferences of diet (Henry et al., 2012; Ledogar et al., 2016).

Hominins as a group are transposed above the Great Apes, but they do so for different reasons. *Homo* does not overlap with apes at all in jaw length, whereas the jaws of australopiths are always stronger than those of similarly sized apes. The common slope for the sample is 0.82; for the dimensionally equivalent space shown, this means that relative strength declines in longer jaws. This might invite speculation that physiological stresses are nominally higher in larger jaws, but the negatively allometric slope here is driven primarily by a highly derived taxon at the lower extreme of the range (modern *Homo*). Modern human mandibles are quite strong for their size. This is an ironic finding, given the universality of cooking with its attendant relief of challenging masticatory effort. Arguing for any type of functional equivalence in this example is apt to be tortuous.

Finally, strength can be assessed relative to body size (estimated here for the fossils and averaged for the extant sample, Figure 2.6). The overall weakness of the relationship between jaw strength and body mass was anticipated by both Hylander (1985) and R. Smith (1993), but for different reasons. Hylander surmised that skulls are not necessarily tuned to gravitational loads in the same fashion as limb bones, while R. Smith expected that correlated effects of somatic size differences would be likely to swamp a purely local biomechanical signal. The message of this relationship for paleoanthropology, however, is that explanations for australopith morphology do not hinge on an assessment of body size allometry. Pilbeam and Gould's (1974) claim that australopiths are best viewed as scaled versions of the "same" animal makes no sense in light of McHenry's (1988, 1991) body-size estimates—there is no allometry to subtract out.

Allometric analysis of biomechanical systems provides a useful comparative yardstick for understanding size-associated changes in efficiency and performance. For example, the long-standing question of whether muscle force scales isometrically or allometrically (cf. Cachel, 1984; Anapol et al., 2008; Perry and Wall, 2008, A. Taylor et al., 2015) is crucial for understanding masticatory performance in fossils (C. F. Ross et al., 2005; Strait et al., 2010; Wroe et al., 2010). The choice of variables in allometric analysis is informed by specific biomechanical or physiological hypotheses, which simultaneously inform and restrain interpretation.

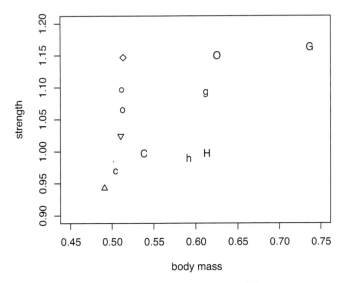

Figure 2.6. Polar moment of inertia$^{0.25}$ plotted against body size$^{0.33}$ (exponents are used to achieve dimensional equivalence). Mean values are plotted for both, since individual body-size values for the hominoid sample were unavailable. Fossil symbols, as in Figures 2.3 and 2.4, are visualized in log$_{10}$ space. For the modern taxa, males are in uppercase letters and females are in lowercase ones. There is no body-size allometry for the australopiths; i.e., the idea that jaw function among australopiths must be understood by a scaling "criterion of subtraction" is invalid. One reason for the criterion of subtraction argument was that body size in australopiths was originally conceived of as a simple correlate of tooth size. This argument was abandoned once body-size estimates were based on postcranial material. Body-size estimates based on McHenry (1992), R. Smith and Jungers (1997), Jungers and Baab (2009), and Holliday et al. (2018); mandibular data from Daegling et al. (2014, 2016).

Total Morphological Pattern

Agreement in paleoanthropology is rare. With respect to taxonomic, phylogenetic, and functional decisions, however, there is consensus that total morphological pattern (Clark, 1950, 1955) is the appropriate arbiter. Its invocation is a signal that the preponderance of evidence is being considered. This perspective is widely credited with resolving the mid-20th century debate over the status of australopiths as hominids or pongids.

As either a theoretical position or a preferred method, total morphological pattern is a relatively undeveloped concept, despite its familiarity. This may be due to the perception that its meaning and prescription are obvious. But from both Clark (1950, 1955) and later advocates, there has been little guidance as to how the concept should be implemented, let alone defined. What the pattern is or how it is recognized is a matter of the "pragmatic test of experience" (Clark, 1955:25), and for most practitioners the theoretical justification is self-evident: bringing more evidence to bear on a research problem is better than having less. With respect to functional analyses, avoiding weighing characters while examining as many of them as possible provides ostensibly more objective conclusions. Clark (1955, 1959) never endorses this view. The idea of measuring everything while weighting nothing is the theoretical justification for numerical taxonomy, which was, of course, never intended as a tool of functional analysis, but rather was intended to serve as an objective classificatory tool or an agnostic phylogenetic approach. In numerical taxonomy, the only legitimate analysis is one in which no metric characters are discarded, owing to the certain pessimism that homoplasy and homology were indistinguishable in most contexts. Implementing this skepticism, however epistemologically noble, is completely unworkable. In the first place, the number of morphological variables that can be imagined in, say, the talus, may not be infinite, but for practical purposes, with the emergence of sliding and semi-landmarks, it is. The second issue is that independence of characters is more a statistical wish than anything biologically tangible.

This analogy to numerical taxonomy paints too bleak a picture. While systematists are required to lose sleep over homoplasy, in assessing a total morphological pattern for functional inference, the status of features as homologies or homoplasies is immaterial, as they are equally susceptible to natural selection.[13] What counts as an independent character in assessing total morphological pattern, however, is not entirely clear. The geometry of an articular facet and its relative size with respect to joint forces can be recruited as separate lines of evidence in a mechanical analysis, but there is no justification, in developmental or evolutionary senses, that these are independent characters. Total morphological pattern, or at least one version of it, coheres with the idea of functional complexes—suites of traits that, together, influence a specified activity. The recognition of modularity in evolution, anticipated in Clark's (1950, 1955, 1959) writings, is reflected in correlation among features within a functional complex. Modularity provides theoretical support for the idea that total morphological pattern can reflect an underlying adaptive reality. At the same time, this presents operational difficulties. A functional module is, as an evolving entity, a single character, as none of the component parts are free to change independently. This should discourage the use of laundry lists of characters to shore up an adaptive, functional, or mechanical scenario in the fossil record. How suites of traits are recognized as a module of evolution follows from examination of patterns of covariation (Cheverud, 1996;

Marroig and Cheverud, 2001). It is not obvious that identification of such modules facilitates functional inferences, even if they underscore the interdependence of traits more transparently. Biological roles still need to be elucidated by reference to nonmorphological (e.g., behavioral and ecological) data.

Even with this admonition, total morphological pattern is not really an approach that gravitates to holistic morphology, despite Tobias's (1985) argument that this should be what the concept should aspire to. In practice, total morphological pattern is a blanket invocation, designed to confer legitimacy to a study, but there is no standard by which a total pattern is evaluated. The only point of agreement is that the concept means more than one character informs a taxonomic or functional conclusion. It is often invoked in descriptions of single skeletal elements (e.g., Churchill et al. 1996; Susman and de Ruiter, 2004). Total morphological pattern is more often than not another name for an appeal to parsimony.

The concept has been amended. R. D. Martin (1968) advocated focusing on "total functional patterns" on the grounds that this provided more explicit focus on adaptation. Lovejoy (1975) and Lauder (1995) further argue that elucidation of "total biomechanical pattern" is more useful in the paleontological setting when behavioral reconstruction was the goal. Even so, Lauder (2003:319) is less than sanguine that this immediately solves issues of implementation, because "the discipline of biomechanics has had a long but relatively superficial flirtation with evolutionary biology." The biomechanical focus is more explicit, however, and provides necessary information for eventual experimental tests of functional hypotheses.

Stern and Susman (1991) suggest that total morphological pattern could be defined by negation. Their juxtaposition of total morphological pattern with "magic traits" in paleoanthropology has a pejorative intent: such characters "tell us everything we wish to know" (Stern and Susman 1991:100). More helpfully, a magic trait is one that provides information about biological roles. It reveals behavior, not just performance. In Stern and Susman's formulation, magic traits in the fossil record have the peculiar characteristic of identifying performance attributes in other traits that were no longer adaptive. In other words, a magic trait goes beyond the lofty goal of identifying biological roles and adaptations: it effectively identifies behavior.

Magic traits were floated in the context of a decades-long debate on early hominin bipedalism that is the subject of Chapter 4. They are distinguished from "diagnostic" traits in the sense that the latter do not rule out particular adaptations as much as they confirm the presence of a particular function or behavior. For example, the genu valgus being indicative of bipedalism is a diagnostic trait. The statement that this feature indicates bipedal locomotion and rules out all other lo-comotor adaptations is more expansive in its explanatory scope and thus qualifies as a magic trait. The extent to which a magic trait is a straw man depends on a lit-eral versus allegorical reading. What is underscored by the concept is that the easi-est inferences proceed from the idea that there is a one-to-one mapping of morpho-logical to behavioral variation. This idea should not be taken seriously, if for no other reason than that observation and experience refute it. Armadillos possess forelimb skeletal proportions that facilitate a fossorial existence, yet this is no rea-son to rule out running as an important behavior in which their limbs participate.

In effect, an insistence on a one-to-one mapping of morphology to behavior ob-viates consideration of biological roles at all. The premise of such roles, as com-plementary to evolved adaptations, is that well-defined morphological traits or complexes can be and are utilized for multiple behaviors. The insidious luxury of paleoanthropology is that the morphology is available for study and hypothesis testing, but the scope of associated behavior is dependent on the initial anatomi-cal interpretation. This leads to peculiar practices. An imagined strict coincidence of morphological features with specific ecological or functional activities can be interpreted as both the inspiration *for* and tests *of* adaptive hypotheses. Meaning-ful tests require independent assessments of activity and behavior. Since this is invisible paleontologically, some surrogate means of inference is required, usually in the form of neontological analogies. These are explored in Chapter 3.

Developmental Perspectives on Bone Morphology

For much of the history of paleoanthropology, morphological traits have been conceptualized, to some degree unconsciously, as static features of the organism that interface with an environment at a specific point in time. In the pursuit of uncovering crucial adaptations, the implicit assumption has often been that these traits are fixed products of the genome. Under this view, variation in trait expres-sion is a nuisance, unwelcome noise for those seeking a functional signal. A more hopeful perspective is that genomic interaction with the environment over devel-opmental time follows some set of morphogenetic algorithms. These processes are recognized in terms of the concepts of functional or physiological adaptation, phenotypic plasticity, and norms of reaction. While these concepts are not strictly interchangeable, all hold that adult morphological variance is created by more or less deterministic processes. At one or more levels of analysis, these processes are mechanically mediated. The important implication for the morphologist is that specificity of activity has some role in determining phenotype.

A succinct definition of evolution is that of a change in allele frequencies over time. There is little utility in this definition for the functional morphologist, however. The reality of developmental plasticity encourages a more explicit consideration of ontogeny for reconstructing evolution. There is nothing novel in this idea, since 19th-century embryologists understood that morphological differences across species could be explained as changes in developmental rate and timing (Gould, 1977). Nor does this recognition, on its own, dictate whether morphology should operate within a functionalist or a structuralist paradigm. The rise of theoretical morphology (Raup and Michelson, 1965; Raup, 1966; McGhee, 1999) provided fuel for the concepts that Gould (1977) and others would recruit to hammer the adaptationist paradigm: contingency and constraint. Alberch's (1980:653) treatment of the latter was sobering to functionalist presumptions: "the production of morphological novelties by developmental programs is not random." Consequently, selection does not sample from a continuous morphospace, and developmental contingency is analogous to historical contingency. Gould (1977) viewed historical contingency as stochastic; Alberch (1980), by contrast, understood developmental contingency to be ultimately explicable by morphogenetic algorithms.

Renewed interest in developmental evolutionary biology ("EvoDevo") has naturally followed from the mid-20th century renaissance of ontogenetic research. Ontogenetic allometry has since figured prominently in morphological studies of primate skeletal and dental function (Ravosa and Vinyard, 2002), under the hope that with consideration of heterochrony, a more empirically grounded synthesis of development and evolution could emerge. Sampling limitations impact the scope of such investigations in the fossil record.

A neglected lens through which to interrogate the hominin fossil record for functional inferences is skeletal biology. Quantification of a fossil's relative size, morphometric shape, and structural biomechanical properties are methodological problems that have been amply solved. Determination of the functional meaning and adaptive significance of these variables requires an analytical framework that has a coherent theoretical foundation. In 20th-century paleoanthropology, that foundation was a principle of bone adaptation known as Wolff's Law. Formally articulated by Julius Wolff at the end of the 19th century, it is attributed to the ideas of Wilhelm Roux (O. Pearson and Lieberman, 2004). The inspiration for the law was the cancellous network of the human proximal femur; the trabeculae were observed to align for optimal resistance to assumed postural loads. Lanyon (1974) provided experimental data for this "trajectorial theory" of bone architecture, concluding that peak principal strain directions coincided with the orientation of trabecular bone.[14] This represents the specific articulation of Wolff's Law, but from it emerged a more general form of the law that, in biological anthropology, became the theoretical foundation for skeletal functional morphology. This general articulation is that bone is a self-optimizing material with respect to its mechanical environment. What the optimizing criterion should be is not always clear. It is usually formulated as maximum strength with a minimum of material (Huiskes, 2000), which requires knowledge of the applied loads as a meaningful test. Despite some subtle variations in definition, the overall message is the same: bone is developmentally responsive to the loading environment. With respect to the paleontological record, this can be restated as an interpretive principle: bone mass and geometry provide direct insight into load history. The allure of this principle to functional inference in paleontology is difficult to overstate, because independent determination of the loading environment can be avoided entirely. Trinkaus (1989) applied this idea to infer high activity but foraging inefficiency in Neanderthals, a combination that Sorenson and Leonard (2001) deemed unlikely, because the coincidence of these two attributes in a single taxon would have been thermodynamically unsustainable.

The logic for extending Wolff's Law to cortical bone is straightforward, as the same tissue is involved, the material properties are similar, and drift and repair occur, as in cancellous bone. Including cortical bone under the umbrella of Wolff's Law also brought opportunities for additional tests of its validity. If the law is generally applicable, certain observations follow. First, load histories should be recoverable from cortical bone mass and geometry, provided the investigator has at least a basic knowledge of anatomy and the behavior of the taxa under study. Second, an optimum strength:mass ratio implies that there are stress and strain magnitudes that are ideal: i.e., too little stress suggests too much mass.[15] It follows that the distribution of stresses throughout a bone (the stress field) should be approximately uniform if there is a single optimum. Experimental data emphatically refute these predictions. Where the skeleton has been sampled *in vivo*, strain gradients (spatial changes in strain magnitude) during physiological activity are ubiquitous (Hylander and Johnson, 1992, 2002; Judex et al. 1997; Blob and Biewener, 1999; Lieberman et al., 2003, 2004). In addition, cortical bone is not distributed so as to provide the greatest structural rigidity with respect to the predominant bending loads in long bones (Demes et al., 1998, 2001; I. Wallace et al., 2014). Bertram and Swartz (1991) and O. Pearson and Lieberman (2004) provide additional critiques—principally, that the outcome of metabolic activity in bone is not directed toward a single mechanical optimization criterion. Skedros (2012), in acknowledging the inconsistent findings in experimental and comparative arenas, notes that there need not be a singular objective of bone metabolic activity that has been targeted by selection across the skeleton. He advocates examining applied loads from the perspective of complexity, rather than exclusively from mode (Figure 2.7). This is certainly a more realistic position than the expectation that bending, on its own, explains bone geometry, since bending probably rarely occurs in isolation *in vivo*. Moreover, Skedros (2012) argues that the microstructural anatomy of secondary bone appears to covary in conjunction with variations in the strain field, most convincingly in terms of preferred collagen orientation in secondary osteons. Here again, however, the caveat of accurately modeling the loading details applies, because the strain field is inferred, rather than observed.

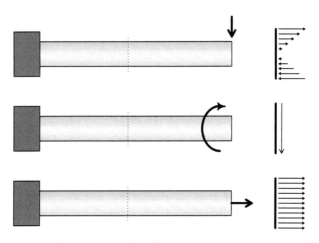

Figure 2.7. The drawings illustrate the modes of loading and their determination of stresses and strains. For visualization purposes, these idealized bones are considered to be fixed on their left ends. *Heavy arrows* on the right ends represent the direction of applied loads. The *dotted line* represents a plane of section, magnified as a *heavy line* (on the *far right*), with the nature of the stress depicted as the *small arrows*. Under bending (*top*), tensile strains decrease linearly from the upper periosteal border to the centroidal ("neutral') axis of the bone, where the longitudinal stresses and strains are zero. Below the neutral axis, compressive strains increase linearly, reaching a maximum at the opposite periosteal surface. By convention, these longitudinal stresses and strains are called "normal" stresses and strains, since they are perpendicular to the face of the cross-section. This contrasts with the condition of torsion (*middle*), in which the cross-section is subject to shear stress. Under an axial load (*bottom*), the normal stresses are similar throughout the section. Analytically, each of these load cases can be examined in isolation, but it should be appreciated that the sources of load probably occur in different combinations under varying behavioral contexts. Normal and shear stresses are present in most loading contexts.

The experimental data do not overturn the general form of Wolff's Law per se, but they rule out an imprint of stress history in bone geometry, given our current understanding. There is incontrovertible evidence that bone is responsive to alterations in mechanical loading (R. B. Martin et al., 2015), but the challenge is to understand how the metabolic algorithms of bone growth, maintenance, and repair have evolved. Research done under the paradigm of Wolff's Law rarely asks the question, What parameter is the target of selection? Historically, the relevant variables to quantify strength-to-mass optimization are not explicitly articulated. Biewener's (1993) hypothesis of a narrow range of safety factors, or Rubin and Lanyon's (1984a) hypothesis of dynamic strain similarity are exceptional in this regard. Dynamic strain similarity is the idea that during vigorous but normal activity, the highest bone strain magnitudes are more or less the same in different animals. It has become increasingly clear that these two "adaptive" principles probably only adhere closely to the mechanics of long bones. For example, contrary to what was generally believed at the close of the 20th century, masticatory activity produces *in vivo* strains significantly below those engendered during vigorous locomotor activity (Ravosa et al., 2010). This means that unless there are large differences in bone material properties between the facial skeleton and elsewhere, safety factors are not constant across bone elements (Figure 2.1). This also undermines the generality of Wolff's Law.

One reason why paleoanthropology has focused on cortical bone mass and geometry for adaptive and behavioral inferences is that the study of cancellous bone was historically impeded by methodological problems of imaging and characterizing the three-dimensional trabecular lattice, as well as the cumbersome nature of the quantitative analysis of that lattice. These problems have been overcome by technological advances in radiographic imaging, as well as automated image analysis. Consequently, the potential for trabecular organization and geometry to enable reliable behavioral inferences is being reexamined in paleoanthropological settings. As is the case for cortical bone, there is no reason to doubt that cancellous bone is sensitive to alterations in load history, owing to support from experimental data (Barak et al., 2011). The current paradigm is essentially identical to that of Wolff's original formulation: trabeculae are preferentially aligned with the principal stresses encountered *in vivo*. This, in turn, leads to the inference that joint load direction and magnitude are imprinted on trabecular morphology in terms of orientation (expressed as degree of anisotropy[16]) and density (expressed as volume fraction[17]), respectively. This logic has been used to argue for modern human–like hip-joint loading in *Australopithecus africanus* and *Paranthropus robustus* femoral heads (Ryan et al., 2018). One observation in their study, which cautioned against using trabecular organization as a diagnostic locomotor trait, was that baboons resemble the hominin pattern more closely than they do that of Great Apes. Ryan et al. (2018) interpreted this as perhaps indicating that the high anisotropy in baboons and bipeds had more to do with stereotypical joint loading than bipedality per se—the idea being that the more isotropic fabrics of apes reflected a more variable locomotor repertoire.

Given the responsiveness of trabecular mass and organization, as well as the documentation of species-specific patterning of cancellous bone (which correspond, to some degree, with locomotor habits), the question arises as to whether Wolff's Law is true, after all. This depends on how literally the law is to be applied. Skedros and Baucom (2007) tested the "trajectorial theory" of cancellous architecture, which posits alignment with principal strains; by definition, this means that trabeculae ought to intersect one another at about 90°. This turned out to be true of artiodactyl calcanei but not of proximal femora of humans and chimpanzees. The hominoid pattern was nonorthogonal with respect to mean intersection angles, in conflict with theoretical expectations. Skedros and Baucom (2007) hypothesized that the observed differences between the skeletal elements were attributable to load complexity: calcanei of sheep and deer function more or less like cantilever beams, and proximal femora do not behave like beams at all (see Chapter 4). The trabecular arrangement in hominoid proximal femora suggests a loading environment dominated by shear.

Taking on Wolff's Law in more general terms, the adherence of trabecular morphology to law-like reflections of behavior is challenged by two important observations. One is that similar arrangements of cancellous bone in the proximal femora of *Otolemur* and *Galago* are belied by distinct locomotor habits, the latter being a more dedicated leaper. Ryan and Ketcham (2002) suggested that similarities of femoral head geometry may condition the nature of loads occurring there in both taxa; their reasoning—that any degree of leaping may compel certain configurations of a trabecular network—is consistent with mechanobiological principles (see the "Mechanobiology" section below). Other exceptions to cancellous architecture and locomotion are no doubt discoverable if we are committed to finding them, especially if "locomotor pattern" is nebulously defined; i.e., vaguely in terms of biomechanical variables. Another critical finding is that the pattern of cancellous architecture in adult bones—those which have been subjected to a lifespan of gravitational loads—may be present in that "adult" arrangement *in utero* (Skedros et al., 2004). This is inexplicable in functionalist terms.

It makes sense that trabecular bone arrangements seem to fit the framework of Wolff's Law best, given that this type of tissue inspired the idea. On the other hand, the process of trabecular modeling and remodeling in the context of changing mechanical conditions is underexplored vis-à-vis comparative studies of cancellous bone patterning under altered stress environments. The outstanding problem is what these "stress environments" amount to in terms of what the trabecular lattice is experiencing. Cancellous bone is usually studied in regions deep to the subchondral bone of synovial joints, in part because this is where trabecular bone is found. Yet we have surprisingly little information about the details of these joint loads; i.e., there is a pronounced asymmetry in knowledge between these forces and the structural details of trabecular morphology. *In vivo* measurements of human hip-joint forces are described as having "nonuniform pressures, with abrupt spatial and temporal gradients" (Hodge et al., 1986:2879). While the claim of "complete documentation" of hip-joint forces in human walking (Bergmann et al., 2001) was probably overstated, technological impediments to a full-field characterization of joint forces are certainly diminishing (e.g., Lenaerts et al., 2008).

The Mechanostat

In the past half century, an articulation of principles of bone mechanobiology within the broader field of skeletal biology has been instrumental for clarifying and critiquing ideas about bone adaptation. This body of experimental and theoretical investigation has sought to explain the evolution of bone tissue, based on the postulate that a large proportion of bone morphological variation is mediated by mechanical forces (R. B. Martin et al., 2015). The goal of this research program, which is pursued in multiple labs around the globe today, is to understand the mechanically mediated algorithms of bone growth, maintenance, and repair. H. Frost (1987, 2003) formulated the idea of the "Mechanostat," which, as the name implies, is analogous to a household thermostat, with bone strain replacing temperature as the variable of interest. Strain magnitude is held to be the switch governing whether bone cell activity is directed toward bone loss, gain, or repair. Specific values (alternatively referred to as "setpoints," "strain thresholds," or "minimum effective strains") determine local activity in bone cell populations. Below a few hundred microstrain ($\mu\varepsilon$),[18] bone resorbs, due to osteoclastic activity uncompensated for by active osteoblasts. Bone loss in astronauts provides a convenient example. The interval between 1,000 and 1,500 $\mu\varepsilon$ constitutes an equilibrium "lazy zone" of bone maintenance. At higher strains, the structural integrity of bone becomes threatened and modeling is induced,[19] usually biased substantially toward osteoblastic activity.

The Mechanostat is elegant in its simplicity. Originally, remodeling and modeling occupied different domains of strain intervals on either side of the equilibrium range (H. Frost, 1987). In this initial formulation, bone was viewed as mostly quiescent in the lazy zone. But with greater appreciation of fatigue risk,[20] as well as the observation of near ubiquity of remodeling activity in some elements, the role of coordinated remodeling activity in bone repair was recognized (H. Frost, 2003). The real value of the Mechanostat is not that high strains require a metabolic response, and low strains can be more safely ignored (this is clear enough from the standpoint of structural integrity). What is particularly useful is the recognition that the osteocyte network has a mechanotransductive capacity: the adaptive response is no longer a nebulous process of optimization.[21]

Despite the heuristic appeal of the Mechanostat, its influence in paleoanthropology has been limited. One reason is that H. Frost's (1987, 2003) formulations were not particularly concerned with load cases on specific bones, but rather with general strain effects on bone tissue. Cortical bone geometry, a true obsession of biological anthropology (Ruff and Runestad, 1992; Lieberman et al., 2004), can be accommodated under the Mechanostat umbrella, but the model does not provide much guidance, other than that high strains need to be met with more mass. The model's disregard of load variation leads to a disinterest in strain mode, which is an essential consideration when contemplating the combined effects of bending, shearing, and axial loads on skeletal elements (Figure 2.7).

Paleoanthropology's general ambivalence is also due to empirical developments that have eroded the Mechanostat's explanatory scope. These include studies indicating that very low strains can induce positive changes in bone mass (Ozcivici et al., 2010); remodeling can be expected under "equilibrium" strains (Bouvier and Hylander, 1996); strain rate (the product of load frequency and magnitude) is a better predictor of bone metabolic activity than magnitude alone (Burr et al., 2002); and recognition that bone has endocrine functions, which may conflict with its mechanical ones (N. Lee et al., 2007; Hamrick and Ferrari, 2008). Collectively, this would seem to suggest that developmental models of bone function have little to offer the paleontologist. This, however, is not the case.

The attraction of developmental models is that once the proximate physiological processes that direct skeletal growth are identified, morphogenetic rules can be proposed and tested. Despite an impressive body of literature demonstrating the sensitivity of bone's structural and material properties to mechanical stimuli, there are as yet no algorithms that successfully predict adult bone size and geometry from a pure criterion of applied stress and strain (an example is provided in Chapter 6). Possible reasons are manifold, but two implications are particularly important: (1) bone form is actually indeterminate from an exclusive reference to the mechanical environment, and/or (2) the mechanical environment cannot be modeled with adequate precision. The latter is almost certain: the boundary conditions that must be specified to produce a description of the stress field (e.g., in a finite element analysis) rely on multiple simplifying assumptions of unknown validity. The former cannot be demonstrated without a fairly complete understanding of the mechanical environment, but it is realistic to think that stochastic factors in the developmental environment introduce noise into morphogenesis. Evolutionary simulation of mechanically mediated bone growth (i.e., a quasi-deterministic model, with modest environmental noise built in) predicts the evolution of nonoptimal phenotypes, in terms of the efficient deployment of cortical bone (Nowlan and Prendergast, 2005). This should not be surprising in the context of evolution, but what is remarkable is that suboptimal results are obtained without the introduction of any additional constraints (developmental or phylogenetic) in the algorithm itself. The implications of this for skeletal evolution are quite profound, suggesting that natural selection is not adept at achieving the most economical solutions for mechanical challenges in evolving skeletons.

Mechanobiology

Since morphology is the outcome of development (and can always be explained as the product of ontogeny), an understanding of variation in fossil assemblages stands to benefit from a mechanobiological perspective. With the demise of Wolff's law and the Mechanostat as deterministic models in all but the most general sense, what does this perspective offer? It is important to appreciate that although mechanobiological approaches for explaining bone morphology date to the late 20th century (see Carter and Beaupré, 2007), this is still a nascent field of inquiry, and there is considerable uncertainty regarding "morphogenetic rules" that the methods seek out. Experiments provide guidance as to what such rules may entail, but generalizations of these rules across contexts (variables such as age, sex, species, skeletal element, or even regions of an element) have met with limited success. Computational modeling provides needed insight into how and whether such rules can evolve, but it also relies on idealizations of processes that rarely yield virtual bones that closely resemble actual ones. Ironically, we were far more certain about how bones evolved under Wolff's Law and the Mechanostat than we are today. But we were also wrong in most details.

Recognition of the ecological and physiological impacts of body mass variation has had an overarching influence on skeletal biomechanics applications in primate and human evolution. Gravitational loads have been held up as the essential consideration for understanding long bone scaling. Whether conceived of as beams or columns,[22] long bones' primary function is to accommodate the problem of weight while still enabling locomotion. Gravitational loads do not act in isolation with respect to a functioning skeleton, since muscular forces regulate postural or locomotor activity and will also impart stresses to bones. Whether gravitational loads or those arising from muscular activity are primarily responsible for conditioning the strain field is a source of controversy (cf. Judex and Carlson, 2009; Robling, 2009). Muscular activity during locomotion is dynamic, and it can be shown to also be so during apparently "static" postural activity (Rubin et al., 1992). The gravitational field represents a ubiquitous source of static load, but this is not a potent mechanical source for facilitating bone growth and development. Bone responds negatively to static loads, as indexed by suppressed formation or active resorption (Robling et al., 2001), and some experiments suggest either that static loads are more or less invisible to bone cells, or that the static stimulus is ignored (Rubin and Lanyon, 1984b, 1987; Lanyon, 1992). Consequently, dynamic strains appear to be the osteogenic mechanical signal.

The reality of skeletal fracture tells us that even if bone strength is the target of natural selection in the skeleton, ultimate strains are still experienced. The presence of safety factors in bone—where peak physiological strains are less than yield strains by a factor of 2 or more—suggests that natural selection has built in some insurance for the skeleton against the odd accident (Ruff, 2006). There are a few reasons to be skeptical of the proposition that fracture avoidance completely explains safety factors. One is that accidents, by definition, are stochastic. If safety factors are adaptations to mitigate fracture risk, fracture events must cull afflicted individuals with a nontrivial degree of frequency. Primates survive fracture events with some regularity, although this, by itself, does not resolve the question (Lovell, 1991). What is clear enough is that natural selection has not eliminated the risk. The cost of that effort would be prohibitive. For example, a uniform safety factor of 10 throughout the skeleton would presumably reduce rates of fracture, but the added energetic burden in locomotion and foraging certainly would represent an intolerable reduction in fitness. The fact that there is an efficient physiological response to fracture, however, suggests that natural selection has favored repair over prevention. The prevalence of bone remodeling throughout the lifespan supports this idea.

If safety factors have not been selected to reduce fracture risk due to accidents, how are they to be explained? As noted above, one possibility is reduction in the risk of fatigue failure. Fatigue can be understood most simply as the formation, accumulation, and growth of cracks in the bone matrix. Thus there are three general strategies for extending fatigue life: stop cracks from forming, allow them to form and discourage their growth, or fix them. One tactic for preventing crack formation is to have sufficient bone mass, so that stresses and the attendant strains never reach a magnitude that facilitates crack initiation. This only revisits the fool's errand of trying to prevent whole bone fractures altogether by means of an unwieldy and metabolically expensive skeleton. Skedros (2012) also suggests that the strategy of structural reinforcement through added mass is a poor one, under the idea that large volumes of material are more likely to contain design flaws than small ones (the "stressed volume effect"). A material tactic for limiting crack growth is to lessen the stiffness of the bone matrix itself by reducing mineralization. This works because strain energy deforms the compliant matrix and there is less energy available to drive an existing crack further. This tactic has non-negligible performance costs, because more muscular energy goes into deforming a bone than into producing movement or doing other useful work.

To some limited extent, bone uses these structural and material tactics, but the most effective option is repair. Remodeling involves the replacement of old bone with new, through the sequential and coordinated activation of osteoclasts (responsible for dissolving existing bone matrix) and osteoblasts (responsible for depositing new bone matrix in the wake of osteoclastic activity). Cracks are removed through a process of targeted remodeling. Cracks disrupt the strain, or fluid flow, in the osteocyte network's canaliculi, which, in turn, initiates a signaling cascade that activates the remodeling process. The presence of secondary osteons thus identifies a locality of repair. There is also some amount of "nontargeted" remodeling that occurs as well, which will produce an identical histological signature. The nontargeted process is thought to be related to systemic requirements of mineral homeostasis (calcium and phosphate) and has no necessary correlation with local mechanical conditions. The relative amount of these two forms of remodeling is uncertain, although R. B. Martin (2002) offered a probabilistic argument that nearly all remodeling is spatially associated with microcracks. If true, then mineral homeostasis could largely be accomplished incidentally, through the process of bone repair.

Peak maximum strains have been treated as the most important variable for understanding skeletal adaptation for several reasons: (1) they allow calculation of safety factors, (2) they are critical in determining fatigue life, and (3) they can be decisive for understanding whether a bone is adapted for structural load bearing versus other functions. The fundamental assumption of the Mechanostat is that certain strain thresholds will evoke a metabolic response in bone. Conversely, it leads to the conclusion that small-magnitude strains have little influence on morphology. Paleoanthropology has operated from this premise whenever the focus is on bone strength. Its generality, however, is questionable. There is abundant experimental evidence that bone maintenance and apposition can occur at strains well below the 1,000 $\mu\varepsilon$ setpoint envisioned by H. Frost (2003), provided that these strains occur at relatively high frequencies (Rubin and Lanyon, 1987; Rubin et al., 1992, 2001, 2004). While reaffirming the sensitivity of bone to a dynamic loading environment, this implicates a wide range of activities as being important factors in osteogenesis. In other words, peak maximum strains represent only one aspect of the mechanical environment to which the skeleton is responsive (Figure 2.8).

Various experiments suggest that both very few ($N < 10$) loading cycles inducing high strains and tens of thousands of cycles on the order of 10 $\mu\varepsilon$ are sufficient to maintain bone mass. The osteogenic potential of the high-frequency, low-magnitude strains is disputed, however (see Rubin et al., 2001; Ozcivici et al., 2010; Kotiya et al., 2011; Kiel et al., 2010), and part of that debate concerns mechanotransduction at such small strains (Turner et al., 1995; Garman et al., 2007; Ozcivici et al., 2010). Deformation of the bone matrix is unlikely to be sufficient for this purpose. The inconsistency of experimental results is undoubtedly due in large measure to the use of different animal models, as well as to distinct clinical populations, but in some contexts, there is a clear association between high-frequency, low-magnitude strains and an osteogenic response.

A related issue is what, exactly, the osteocyte network is monitoring. Burr et al. (2002) noted that the interaction of load intensity and frequency could be collapsed into a single variable, strain rate (units $= \mu\varepsilon/s$). This is not just analytically convenient for the morphologist, as it also provides a variable that can be monitored by the osteocyte network. One intriguing property of cellular monitoring of the strain environment is that, after the onset of activity, mechanical signals are ignored after a brief period (Robling et al., 2001; Turner and Robling, 2003). Thus the osteoregulatory signals are not accumulating as a simple function of the number of cycles experienced on a daily basis (Figure 2.9). Therefore, loading history is not recorded as much as it is sampled. This makes sense from the perspective

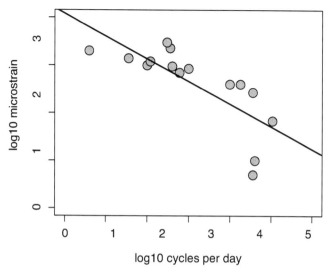

Figure 2.8. The maintenance of bone mass is related to the interaction of peak strain magnitude and the daily number of loading cycles. The *circular points* are drawn from experimental studies in which different combinations of load magnitude and cycles were shown to be sufficient for maintaining, augmenting, or slowing loss of bone mass in treatment subjects. The best-fit line represents a hypothesis that various combinations of strain magnitude and frequency will evoke a modeling response. Considered collectively, the experimental data indicate that a few high loads experienced daily may induce the same response as multiple cycles of low strain. The applicability of this principle across species, skeletal elements, age, and sex is unknown but will be variable. Data for points from Rubin and Lanyon (1984b, 1985, 1987), McLeod and Rubin (1992), Rubin et al. (1992, 2001, 2002, 2006), Qin et al. (1998), Cullen et al. (2001), Hsieh et al. (2001), and Ozcivici et al. (2010).

of metabolic activity, since there are clear physiological limits to the amount of bone that can be added or replaced on a daily basis; i.e., in the event of daily overloads, there is no metabolic solution that will stave off failure in the long term.

There are also age effects with respect to skeletal plasticity that have implications for functional inference. Skeletal senescence is a persistent public health concern, and the molecular and cellular mechanisms underlying it are becoming understood (Burr and Allen, 2014). In terms of metabolic processes, the aging skeleton is less responsive to mechanical stimuli, and the tight coordination of bone removal and apposition that is critical to the remodeling process can be disrupted by a variety of factors. Exercise-induced benefits, in terms of mass and mineral content, are transient in older individuals if a training regimen is discontinued (Forwood and Burr, 1993). The magnitude of bone response to changing mechanical environments shows a general decline with age, with adults exhibiting diminished

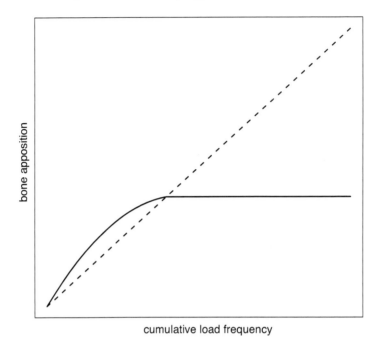

bone apposition

cumulative load frequency

Figure 2.9. The assumption of a perfect relationship between mechanical stimulus and bone modeling or remodeling activity is contradicted by empirical investigations. The *dashed line* represents an expectation that the number of loads experienced over the course of a day eventually results in a proportional increase in bone mass. What actually occurs is more accurately depicted by the *solid line*: after a certain number of loading cycles the osteogenic signals cease, and no further addition of bone mass will result. This, too, is overly simplistic, because the spacing of loading events also influences the sensitivity of the osteocyte network (see Figures 3.6 and 3.7). The events of the *x* and *y* axes are asynchronous: the loading frequency is depicted over a single day, whereas the process of bone apposition will take weeks to complete.

capacity to add bone mass, compared with juveniles (Kannus et al., 1995). One finding of particular interest is that vigorous exercise regimens prior to and through the adolescent growth spurt appear to be a potent factor for ensuring good skeletal health through adulthood (Nordstrom et al., 2005; Warden and Fuchs, 2009; Gunter et al., 2012). The implication for interpretation of skeletal remains is that insofar as the skeleton provides a record of the mechanical environment over the lifespan, the functional-mechanical signal in adult skeletal morphology may be more directly related to adolescent, rather than perimortem, activity.

One of the puzzles facing mechanobiology is specifying the origin of mechanical causality in ontogeny. As this approach assigns an overarching role to the strain field for directing skeletal development, the question arises as to how the appearance of species' differences in morphology occurs so early in development. For example, African apes, despite their close phylogenetic affinity, display mandibles that are distinct morphologically in infancy, before dietary preferences and feeding behavior have had time to shape morphology (Daegling, 1996). A reflexive explanation is to invoke species' genetic differences. No doubt there is truth to this, but it is particularly unhelpful to the functional morphologist, except for providing an escape clause for a failed hypothesis. The *in utero* mechanical environment has a large role to play, and Carter and Beaupré (2007) reviewed convincing theoretical models that predicted sites of primary and secondary ossification from purely mechanical criteria. But this only kicks the puzzle to an earlier stage of development. Invoking K. Vogel (1991), one can allow the premise that all development involves physical causes. Thus mechanics is always operative, only at different scales of analysis. This does little to resolve the question in practical terms, however. Much remains to be learned, even if a general epigenetic explanation for morphogenesis is essentially correct. Yet it is indisputable that species' differences in skeletal morphology are not entirely dependent on postnatal activity.

Interpreting Bone Morphology through Phenotypic Plasticity

The dynamic response of bone to the mechanical environment, and the apparent peak of that responsiveness during adolescence, indicates that investigation of phenotypic plasticity deserves greater attention. Developmental lability in trait expression obviously has a heritable component, but one cannot argue that each of the emergent polymorphisms in a population represents an adaptation by strict historical definition. This is not a crucial *biological* problem, but an *ontological* one: what matters is whether phenotypically plastic traits are enhancing fitness in the short term. Intraspecific variation is often treated as a nuisance for functional inference, but the bounds of phenotypic plasticity are important to discover, in order to underscore the difference between historical and ahistorical (i.e., evolutionarily "invalid") adaptation. The assumption that underlies comparative morphological analysis is that these bounds are relatively restricted and do not match the magnitude of interspecific differences. Experimental evidence, however, shows that, at least in some cases, this assumption does not hold (Ravosa et al., 2016).

The concept of phenotypic plasticity does not indicate a return to Lamarckian thinking. There is no need for vitalism to serve as an explanation today; the material basis for the mechanisms were understood in general terms with the early

20th-century Modern Synthesis, and specific examples have accumulated since (Via and Lande, 1985; Pigliucci, 2001). It is clear that there are genetic under-pinnings to plasticity, even if many details are as yet unknown. But the idea that ontogenetically mediated morphological outcomes are heritable appears to reduce the role of adaptation (by its historical definition) in explaining morphological evo-lution. This is not so much a problem for evolutionary theory as it is an indicator of the hardening of the adaptation concept in the effort to standardize methods of adaptive and functional inference in paleontology.

What is crucial is resolving how phenotypic plasticity informs the relationship of functional morphology to evolution. Even with a recognition that the evolution of development essentially *is* morphological evolution, current practice in paleo-anthropology is to treat traits in essentialist rather than processual terms; i.e., as fixed things on which selection may act. It would be a mistake to argue that phe-notypic plasticity satisfies a nonhistorical definition of adaptation (i.e., a proxi-mate manifestation). Not all plastic features are adaptive, and many instances of reaction norms are deleterious. One skeletal example is provided by osteophytes, in association with arthritic joints.

Phenotypic plasticity encompasses distinct concepts, each with subtle implica-tions for interpreting morphology with respect to biological roles. Cogent re-views have been provided by B. Hall (2001), Crispo (2007), and Ghalambor et al. (2007). Genetic assimilation and the Baldwin effect both explain selection of adap-tive phenotypic changes in response to proximate environmental stimuli. The difference between these two processes lies in assumptions about how the initially altered phenotypes are ultimately encoded in the genome. The Baldwin effect re-lies on fortuitous new mutations to fix traits by making them more concretely heritable. In contrast, genetic assimilation assumes that there is existing genetic variation—unexpressed phenotypically—that can be selected by proxy once the plastic response is successfully operating in the environment. In neither of these cases is the initial phenotypic manifestation magically sequestered by the genome in classic Lamarckian fashion. Instead, phenotypic plasticity provides an oppor-tunity for sustained selection, as long as the latent genetic foundation is present.

With the ongoing focus of the bone mechanics literature on the developmen-tal sensitivity of morphology with respect to biomechanical variables, a more ideal advertisement for the relevance of phenotypic plasticity for understanding skele-tal evolution is hard to imagine. The processual linkage of developmental variation with adaptation (in the historical sense) is a matter of recognizing epigenetic sources of variation as a catalyst for species-level morphological variation, molded by natural selection. Developmental variation is not limitless but nevertheless provides evolutionary opportunities.

Reaction norms themselves are heritable, and they are finite in their effect. The evolvability of skeletal morphology is not assumed to be limitless, but there is a supposition that the potential range of plastic responses is equivalent across most comparative contexts. This is an understandable position from the standpoint of expedience, but it does not have much more than hope supporting it. Genetic background has a significant effect on the responsiveness of developing bone (Robling and Turner, 2002; Judex et al., 2004; I. Wallace et al., 2012), and no doubt the inconclusive or contradictory findings across studies are due to an inability to control for these factors. This is particularly depressing for the paleontologist, who has little access to genetic information beyond phylogenetic concoctions.

Until this point, phenotypic plasticity has been discussed in a nebulous fashion. How does this phenomenon manifest in bone, and how can this information apply to the fossil record, if at all? Figure 2.10 explores the challenges of applying the implications of this real and potent phenomenon to paleontological contexts. Following Pigliucci (2001), units A–D in each graph represent distinct genotypes within a taxon. These should not be considered identical to species designations, because interspecific contrasts are problematic in the context of thinking about evolvability within populations. By assuming uniformity of plasticity within a taxon, we retreat into an essentialist mindset—thus defeating one of the primary reasons to study plasticity in the first place. The *y*-axis is some aspect of bone morphology: this can be cortical geometry, degree of mineralization, anisotropy, or simple indices of mechanical performance. Load history (the *x*-axis) is some external mechanical variable known to have influence on the *y* variable. This could be represented by such variables as strain rate or a daily stress stimulus.

In the case of flat reaction norms but no plasticity, each genotype has a morphological response that, from the perspective of load history, is predetermined. Ontogeny has no role in conditioning morphology. If A–D are species, their transpositions could be seen as historical adaptations, given ecological circumstances that dictated different fixed performance capacities. The particulars of the immediate mechanical environment are irrelevant.

The "paleontologist's dream" model shows that there is a single predictable relationship between the mechanical environment and skeletal morphology, such that the form of bones provides a faithful account of an individual's load history. Here the environment determines morphology and is logically inferred from it. Another way to state this is that the phenotype is buffered from genetic variants. Implicitly,

Reaction norms, no plasticity

Figure 2.10. Under uniformitarian assumptions, reaction norms and phenotypic plasticity characterize populations in the fossil record. Their reality is bound to create problems of interpretation. In this example, the phenotypic response is some feature of bone morphology (e.g., changes in relative bone mass), and the environment is some aspect of load history (e.g., cumulative stress over a lifespan). Consider four genotypes: A, B, C, and D. In the condition *Reaction norms, no plasticity*, each genotype is associated with a particular morphology, and variation in load history is completely impotent. There is no reason to believe this is realistic. The *Skeletal biologist's nightmare* depicts a genotype–environment interaction that produces divergent responses to load history, at least half of which would appear to be highly deleterious. As the mean responses of each genotype across environments are equal, there is no "net" plasticity in this example. This state of affairs, fortunately, does not have any credible data to support it. A completely different situation, but of equal concern in the paleontological context, is the *Paleoecologist's nightmare*. In this instance, there is no plasticity or any reaction norms. This has the superficially attractive property of enabling the researcher to completely ignore a variation in load history, but it also implies that the phenotype is completely canalized. At certain scales of analysis, this relationship has some validity (e.g., femora are never confused with calcanei), but it does not explain, for example, variation in bone mass and geometry within elements. *Differential plasticity*, in contrast, may succeed in explaining some of this variance. In this situation, the principal difference among genotypes is in the degree of plasticity; reactivity to load history is variable. Much of the data in the experimental literature is probably explicable by this kind of relationship. The *Paleontologist's dream* is that there is a single pattern of plasticity that can be described across all genotypes. The implication that there is a singular norm of reaction means that activity patterns can theoretically be retrodicted from morphology. When it is argued that load history is reflected in bone morphology, this is the scenario that is invoked. Of the possibilities depicted within Figure 2.10, the *Functional morphologist's nightmare* is probably the relationship that approximates reality, given comparative and experimental data considered in the aggregate. This envisions variable plasticity and distinct norms of reaction. The nature of this patterning has yet to be established but is certainly discoverable. Finally, the *Panglossian hope* presents the case where plasticity is identical across distinct norms of reaction. Sensitivity to load history is completely predictable, and what might be termed a "baseline" morphological condition is determined independently. If "species" is substituted for "genotype" here, this depiction reinforces essentialist concepts, while plasticity, although present, presents no special interpretive problems. Based loosely on Pigliucci (2001), who adapted it from Via (1987).

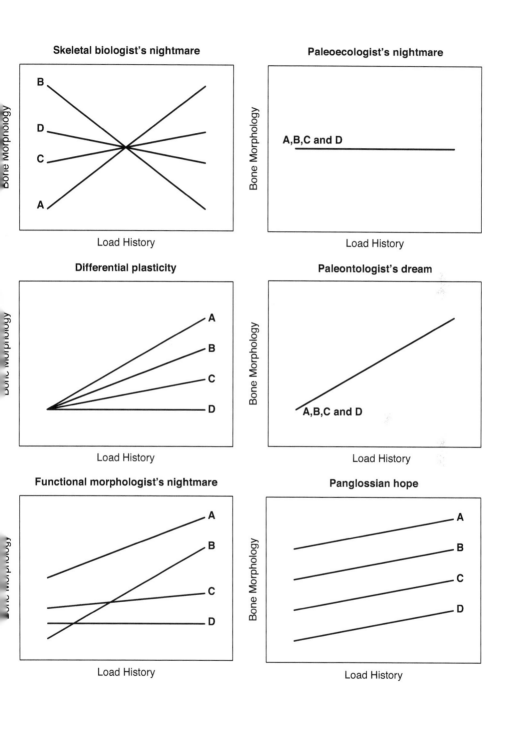

this is the form of many narratives regarding the significance of skeletal morphology in human evolution: activity makes the bone. Ironically, this narrative is often invoked in cases where adaptation is conceived of in rigidly historical terms: selection has found a singular morphology for a particular mechanical environment. But plasticity is really the "adaptation" that matters in this case, and the capacity of bone to tune itself in step with environmental variation is what has evolved.

If bone plasticity vis-à-vis loading history is not conserved, a provocative "differential plasticity" model applies. The reaction norms are created by varying plasticity across units. Unit A is highly reactive to load history, whereas D is completely insensitive. With respect to bone tissue, this case of canalization is the one that begs for an explanation. From a mechanobiological point of view, why would it be beneficial to lose the capacity for adjustment? If we shift perspective and consider the units to be particular bone elements within an organism, this model makes sense both empirically and theoretically. Units A–C represent bones that have different ontogenetic responses to loading variation, which corresponds to what is thought to occur in different regions of the skull (Wood and Lieberman, 2001; Daegling, 2004). Unit D, by contrast, represents elements that experience very little variance in load developmentally, such as the auditory ossicles. This is not a case of phenotypic plasticity, however.

There are also at least three theoretical nightmare scenarios of skeletal plasticity facing paleobiologists. The "paleoecologist's nightmare" is that bones, and therefore the fossils they have become, simply do not react to the loading environment, nor do they display any norms of reaction. There is no reason to believe that either postulate is remotely true. The "skeletal biologist's nightmare" is one in which all imaginable plasticities are potentially operable, given appropriate genetic variation, although, as Pigliucci (2001) notes, this graph implies that there is no *net* genetic variation. This scenario is implausible, since it presupposes that completely distinct morphological solutions to a particular mechanical circumstance (a matter of physics) can function equally well (and are developmentally obtainable). Finally, the "functional morphologist's nightmare" is one in which distinct norms of reaction are associated with particular plasticities, but not in any recognizable pattern. This is drawn here as a worst-case scenario, the details of which are unlikely, based on our current understanding of skeletal biology.

The "Panglossian hope" graph overcomes this problem by having similar plasticities embedded in distinct norms of reaction. It is an attractive model, in that it accounts for bone's capacity to adjust to environmental variation. This is realistic to some degree, except that what remains largely unknown is whether the lines are indeed parallel—i.e., is the nature of plasticity vis-à-vis loading parameters more or less the same across the board?

There is no reasonable expectation that we will ever accumulate a fossil inventory of hominins that will permit evaluation of these alternative models, and, in any case, such inference would require a reliable accounting of load histories of the fossils themselves. The value of investigating these alternatives, however, is that they can provide guidance about the nature of plasticity with respect to, for example, locomotor or masticatory morphology. Linking genotypic and phenotypic variation is not merely aspirational at this point, so moving from the observation that morphospace is discontinuous to an explanation of why that occurs represents an unanticipated avenue for understanding variation in the fossil record.

Phenotypic plasticity is obviously an essential consideration in paleobiology, but the problem is that it is analytically inaccessible, except in terms of statements of statistical likelihood from comparative data (see Nunn, 2011). It is, in a sense, an inoperable paradigm. The twin paradigms of Wolff's Law and the Mechanostat are also problematic, so the pertinent question is, What has replaced them? No convincing umbrella theories present themselves, which is attributable to a growing realization that a universally applicable algorithm of bone growth and adaptation does not appear to be realistic, given current knowledge. There are, however, a number of established principles that are of use to the functional morphologist. These follow from the general premises of both Wolff's law and the Mechanostat, but they are more restrictive. Eight such principles are listed below and should not be considered exhaustive, though all are consequential for functional inferences.

First, bone is responsive to its mechanical environment, and that responsiveness depends on dynamic signals that are ultimately generated during normal physiological activity. Second, bone's adaptive goal over the lifespan may be as much avoidance of fatigue failure as it is prevention of acute failure, which, in a stochastic environment, is probably impossible. Third, given that bone's mechanosensitivity is only operative over brief intervals, load is merely sampled (rather than continuously monitored) by the osteocyte network, meaning that rare bouts of activity are less likely to be detected and, thereby, are potentially ignored in shaping morphology. Fifth, strain rate—the product of load frequency and strain magnitude—is presently the best candidate variable for a universal osteogenic signal. What is not known is whether there are components of the strain tensor that are particularly potent in driving metabolic responses.[23] Skedros (2012) argues that shear components are good candidate variables, in part due to bone's relative weakness under this type of strain. Sixth, bone metabolic activity with respect to modeling is targeting some goal other than the minimization or homogenization of strain. As yet, it is not clear whether this is due to physiological constraints or to bone strength being of secondary importance relative to other selective parameters, including performance variables. Seventh, the metabolic strategy for long-term bone health appears to be repair (secondary remodeling), rather than prevention (or even avoidance) of damage. We would expect different safety factors in bone if the remodeling response is generally efficient. Eighth, the recording of load history in terms of bone mass and geometry is—at best—indirect, and the reflection of activity is most salient from the adolescent stage of life history for primates.

What has largely been unappreciated is that descriptions of one aspect of skeletal adaptation in the paleoanthropological literature rely on a mechanobiological narrative to explain it. Load management—the structural and material responses to experienced stress and strain—invokes physiological adaptation over historical adaptation as an explanation for the emergence of bone mechanical properties, such as strength and stiffness. This is an intentional shorthand means of linking morphology to the mechanical environment (Ruff et al., 2006).

Teleonomy Reexamined

Williams (1966) advocated the development of a specific field of inquiry, teleonomy, to bring methodological rigor to the study of adaptation. His principal objection to contemporary practice was that adaptation could be invoked for any behavior or trait, as long as a believable accompanying narrative of natural selection could be articulated. This form of argument is frequently encountered in paleoanthropology (Tattersall, 2002). Empirical falsification of such proposals is difficult and is almost always indirect (i.e., relying on some form of contrary data, independent of fossil morphology). A science of teleonomy, Williams (1966) reasoned, could resolve the problem of distinguishing adaptation arising from selection from the fortuitous effects of that process. For example, it is invalid to argue that a terrestrial organism's possession of mass represent an adaptation for remaining bound to its locomotor substrate. This discrimination of effects versus adaptation was foundational for the critiques of adaptationism that followed (Gould and Lewontin, 1979; Vrba, 1980; Gould and Vrba, 1982).

Despite Williams's (1966) demonstration that discussions of adaptation were frequently undisciplined (or even incoherent), the call for a new field of study has apparently gone unheeded. There are no graduate programs in teleonomy or institutions for teleonomic research. The literature in evolutionary biology still has its major focus on adaptation, but there is scarcely a mention of teleonomy. This could be explained in one of two ways: either the concerns raised by adaptationist critiques have been overstated, or a discipline of teleonomy cannot stand on its own and is impractical. There is some truth to both positions. A restrictive definition of adaptation that has been generally embraced—applicable only to traits arising from the process of natural selection—prescribes a fairly narrow practice of matching functions with selective agents (Curio, 1973). This does not mean that teleonomic practice would not incidentally identify biological roles or exaptations, but if the objective of functional inference is to reconstruct behavior in a paleo-anthropological context, then the focal concept is aptations in general, not adaptations in particular. Fisher (1985) advocated a pluralistic approach to morphological analysis, where adaptation is just one part of a "causal plexus." Table 2.1 provides a synopsis of various endeavors of morphological inquiry. It is admittedly subjective and idiosyncratic, but my purpose in presenting it is to underscore the position that different approaches are trying to elucidate separate things, despite the fact that all have more than a trivial interest in uncovering adaptation.

One consequence of the adaptationist critique was the de-emphasis on natural selection and the concomitant elevation of historical contingency and morphogenetic constraint as important drivers of evolutionary change. This can lead to the conceptualization of morphological evolution as a zero-sum game (adaptation *versus* constraint, not adaptation *and* constraint). If constraint had been underemphasized, then it followed that adaptation's role in evolution was exaggerated. The more reasonable view is that constraint may limit the variety of adaptations but not necessarily their pervasiveness. This is recognized, if not fully operationalized, in the fields of constructional and theoretical morphology.

Lauder's (1981, 1982, 1995) practical conception of evolutionary morphology arguably obviated teleonomy in paleontology, because of its formalization of a criterion for discovery of historical adaptation. Another interpretation is simply that evolutionary morphology is teleonomy under a different moniker. This is unlikely, on the grounds that theoretical concerns raised by advocates of teleonomy are not the same issues expressed in evolutionary morphology. These issues are focused on the legitimacy of the concept, as distinct from teleology (Pross, 2005; Auletta, 2011). If teleonomy stands on its own, the role of the organism is shifted from a passive object of evolution to an agent of its own evolutionary fate. The logic is not obtuse. The species niche is determined by the activity of the organism, so the ecological context and, ultimately, evolutionary fate are thus self-defined. This har-

Table 2.1 Disciplinary coverage of functional inference

Field of inquiry	Scale of analysis[1]	Phylogenetic considerations[2]	Developmental considerations[3]	Object of discovery[4]	Application in paleoanthropology
Biomechanics	tissue to modular	post hoc, if any	not required	performance/integrity	ubiquitous
Functional morphology	modular to organismal	post hoc, if any	not required	aptations	ubiquitous
Evolutionary morphology	modular to organismal	central	implicit	historical adaptation	common
Ecomorphology	modular to community	implicit	implicit	fitness/behavior/adaptation	common
Constructional morphology	organismal	explicit, not clearly operational	explicit	constraint	absent[5]
Theoretical morphology	modular to organismal	implicit	central	morphospace (constraint)	absent[5]
Mechanobiology	molecular to modular	implicit	central	morphogenetic algorithms	rare

[1] Possible levels are molecular, tissue, modular (an anatomical complex, e.g., musculoskeletal levers), organismal, population, species, community.

[2] "Post hoc": systematics do not figure into the original analysis. "Central": essential part of an analysis. "Explicit": acknowledged as consequential but may or may not be formally incorporated into an analysis. "Implicit": assumed to be important but not included in an analysis. In practice, any field may utilize phylogenetic data to inform the conclusions, but only evolutionary morphology formalizes the practice.

[3] As with phylogenetic considerations, developmental data can be applied to an analysis, even if it is not required. The terms represent typical practice, and exceptions can be found.

[4] "Aptations" refer to any performance attributes or behaviors that enhance fitness, regardless of their maintenance by selection. Biological roles are more safely categorized as aptations rather than adaptations. All fields generally are concerned with adaptations, even though the term is invoked with disparate intent across different studies.

[5] I am not aware of specific, formal applications to paleoanthropological data, but paleoanthropologists are cognizant of the utility of these perspectives (see Sterelny, 2011; Henke, 2005). The lack of applications has more to do with operationalizing the perspective with fossil data than to any theoretical objections.

kens back to E. Russell's (1916) objections to mechanical reductionism in morphology, on the grounds that the autonomy of the organism was discounted. The undercurrent of vitalism is bound to arise when biological systems are recognized as teleonomic. But the question of whether organisms direct their own evolution is not necessarily mystical in its premise. It is logical, and necessary, for organisms to have evolved "models" of the environment that permit effective information processing (Auletta, 2011), and this fits perfectly well with the structuralist view that organisms will seek out appropriate environments, based on their sensory and motor capacities. A teleonomic view of the organism is only metaphysical if one denies that cognitive and sensory systems can arise through an evolutionary process. Of course, an organism's understanding of its surroundings does not require consciousness. For example, the osteocyte network's monitoring of the strain environment in bone functions as a model of the environment. The crucial question for skeletal biology is what kind of environmental variance is actually perceived and responded to. These topics are revisited in Chapters 6 and 7.

Form versus Function: Philosophically Trivial or Pragmatically Crucial?

In biology, morphology has always been marked by theoretical tensions that are largely variants of the original structuralism versus functionalism dynamic of the 19th century. E. Russell's (1916) holistic outlook was essentially functionalist. The basis of Gould and Lewontin's (1979) attack on the adaptationist paradigm was structuralist, as was the subsequent development of evolutionary morphology.

Theory has been a neglected topic in morphology. No doubt this stems from the fact that it is fairly obvious to us that related animals look alike, and that what animals do well accords with how they are put together. Adaptation endures as the pivotal concept, but pedantry has also been our enemy. What we call a trait, or the functions facilitated by its conformation, do not matter to the organism. Whether a trait is primitive or derived, an adaptation or exaptation is irrelevant to it. To argue otherwise entails the belief that primitive characters are less useful, because of their relative antiquity, or that a facultative capacity is less beneficial, because it has not been directly filtered by selection. The utility of primitive traits does not necessarily vanish following a cladistic event (Fleagle, 1983; Duncan et al., 1994; C. F. Ross et al., 2002).

For paleoanthropology in particular, the historical tension between functionalism and structuralism was brought to light through the critiques of the adaptationist paradigm (Gould and Lewontin, 1979; Gould and Vrba, 1982). The functionalist perspective persists largely because it is consistent with an apparently congruent mapping of morphological and ecological variation. Practically, it seems to work well much of the time. The structuralist perspective does equally well, given a shift in emphasis from environmental to organismal agency. By acknowledging constraint, however nebulously, structuralism explains discontinuous morphospace more satisfactorily. Skeletal variation makes more sense under a structuralist worldview.

Philosophically, morphologists are sympathetic to the idea that a holistic functional morphology is a worthwhile pursuit. The implementation of this idea in practice is completely undeveloped. Pigluicci (2003) argued that concerns over the "integrated" organism (in contrast to character complexes within organisms), which is the presumption of the holistic perspective, is banal at its core. Until there is a neontological demonstration of the usefulness of this perspective, paleoanthropology need not rush to abandon atomistic practices. If the goal of functional morphology is to understand how animals deploy their anatomy to effect "useful" behaviors, the discovery and identification of historical adaptation *sensu stricto* should not be the primary endeavor. Given an interest in reconstructing behavior in the fossil record, biological roles and exaptations are equally important.

Evolution can be conceived of as the historical modification of developmental processes. Skeletal mechanobiology represents an avenue through which the morphogenetic rules of bone development can be applied, even to paleontological data. The scope and pervasiveness of biomechanical influences on development are discoverable, and the challenge is to incorporate developmental variation into models of evolutionary change in populations.

There is now ample evidence that bone morphology is not explicable as the simple outcome of natural selection for efficiency and economy with respect to physical activity. Evolvability of the skeleton is a research perspective that can address questions of why certain regions of morphospace are inaccessible to organisms. For paleoanthropology, this puts us on the path toward operationalizing ideas of constraint into research practice.

Chapter Three

Approaches to Functional Inference in Paleoanthropology

The previous chapter detailed a number of different perspectives on the study of morphology and indicated how they inform or are informed by evolutionary theory. These perspectives are meant to help us understand processes by which morphology changes at various scales. This chapter alights more firmly on praxis, but the reader will notice that the theoretical issues of the previous chapter will insinuate themselves throughout the present one. Ideally, the practice of functional inference should address the important questions posed by theoretical perspectives. In paleoanthropology, this happens some but not all of the time. One of the recurrent themes in paleoanthropology has been our tendency to view novel methodological approaches as capable of addressing more questions than they are actually equipped to do.

A survey of self-identifying evolutionary morphologists would find agreement with the proposition that the principles of functional inference, and the execution of a research program, should be congruent among paleoanthropology, paleontology in general, and neontological investigation. Yet a frequent lament is that paleoanthropology operates by a different set of postulates and standards than organismal biology in general and is worse off for it. Cartmill (1990) believes paleoanthropology is theoretically compromised by the premise that humans are unique relative to the rest of the biosphere and, consequently, cannot be completely understood by reference to it. This supposedly makes human evolution inexplicable, or at least undisciplined in its arguments (Tattersall, 2000). The idea is not problematic with respect to functional inferences based on descriptive bio-mechanical analyses of fossils, because these rely on a comparative framework to evaluate relative performance. By itself, the use of comparison signals that homi-nins can be productively contrasted with other animals in order to learn some-thing (Nunn, 2011). Cartmill's (1990) and Tattersall's (2000) objections apply more to organismal or social scales of analysis with respect to adaptation.

Tattersall (2000) sees the disciplinary endeavor as irrational, arguing that there is something peculiar to the discipline that compromises objectivity. Caspari and Wolpoff (2012) refer to this as the DuBois syndrome, after the Dutch scientist who discovered Indonesian fossils now attributed to *Homo erectus*. Dubois did the un-forgiveable and changed his mind about the significance of his finds, finally set-tling on an interpretation that the fossils represented a very large gibbon. The syndrome bearing his name describes behavior whereby theory conditions interpretation in an outsized way, such that observations are more or less pre-defined by one's epistemological predilections. This is not peculiar to paleoan-thropology, however, but is inherent in human cognitive processes (Taleb, 2007; Kahneman, 2011).

Despite their very different emphases with respect to evolutionary processes, both Caspari and Wolpoff (2012) and Tattersall (2000) envision morphology's crit-ical contributions to be in taxonomy and phylogeny. Resolving the status of fos-sils in these areas answers important questions in human evolution. The role of functional morphology for understanding human evolution, however, is margin-alized. White et al. (2009:86) were atypically transparent in arguing that their de-cision to bestow generic status on a set of remains was meant to "express both phyletic proximity and circumscribed adaptive systems, with ecobehavioral and morphological conditions being integral parts of the latter." In this case, adaptive inferences, informed by functional analyses, are given some role in taxonomic deci-sions. This does not mean that adaptive inference precedes taxonomic decisions in morphological analysis. A taxonomy that emerges solely from morphometric distinctiveness—a statistical decision in the best cases—obviously predetermines subsequent adaptive interpretations (Vrba, 1980; Table 3.1). This is not problematic as long as statistical distinctiveness is adequate to infer meaningful performance differences (i.e., those which would have a tangible effect on fitness). This is sel-dom explored explicitly.

These problems are exacerbated in a paleontological context. It is clear that if morphological distinction is the basis for taxonomic decisions, then the resultant taxonomy will condition interspecific patterns of inferred adaptations. Table 3.1

Table 3.1.   Influence of morphological variation on taxonomic decisions
and adaptive inference

A. Dependence of morphological observation and taxonomic decisions

	Species or other OTU	
Morphology	Exactly 1	>1
Statistical or categorical differences present	examine spatiotemporal dependence of variance	OK
Statistical or categorical differences undetected	OK	paleontologically intractable

B. Adaptive significance of trait distribution, based on taxonomic decisions

	Species or other OTU	
Morphology	Exactly 1	>1
Statistical or categorical differences present	multiple adaptive states or no variance in fitness	species-level adaptation
Statistical or categorical differences undetected	species-level adaptation	universal adaptation or universal constraint

Notes: *Part A* is based on Vrba (1980). "OTU" is Operational Taxonomic Unit and could refer to either sub- or supra-specific entities. The "OK" designation means there is no conflict between a taxonomic decision and patterns of morphological variation. Adaptive significance (*part B*) is intended to signify default inferences based solely on character distribution, as would be the case in the absence of a targeted functional investigation of the biological roles of the trait(s) in question.

outlines this relationship. When morphology is the basis for taxonomy, the case in which someone has two species that are morphologically indistinct is methodologically intractable. In such an instance, the functional morphologist is more or less forced to conclude that the same degree of adaptation is operative, but the fossil anatomy itself may not indicate whether there is necessarily an adaptive signal at all.

The case of having a polymorphic single species can be resolved by detailed examination of spatiotemporal variation. This is a better outcome for the taxonomist, but the functional morphologist again faces an interpretive dilemma: the variance in particular traits could mean adaptation to different ecological conditions across a species' range, but it could also indicate a general absence of selection, at least in terms of directionality across the species' membership as a whole. The two unproblematic cases for the taxonomist are dissimilar morphs corresponding to separate species, and a lack of distinctiveness in an assemblage of

a single species. These two distinct situations will probably yield a singular interpretation from a functional morphology perspective: each situation indicates
species-level adaptations. The challenge for both taxonomy and functional morphology would appear to be how to deal with the "problem" of variation. The issue, viewed as an extension of Vrba's (1980) taxonomic dilemma, seems to be much
more acute for the functional morphologist. In practice, however, functional morphology can deal with the consequences of variation fairly explicitly.

A general approach to adaptive inference is provided in Figure 3.1. In this example for a given skeletal element, there is variance in joint conformation (*J*) as
well as element length (*L*). For any individual, these variables, in combination, provide information on performance (*P*), which will differ within a sample and across

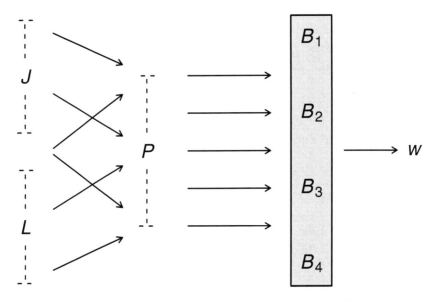

Figure 3.1. The path of inference from morphology to fitness. Anatomical variation is
represented by two characters (*J* and *L*) of a single skeletal element, representing joint
geometry and element length, respectively. The combination of these variables in any
individual has some nontrivial impact on biomechanical performance (*P*). Performance
attributes are involved in the conduct of biological roles (*B*, here arbitrarily numbering 4),
which may or not represent historical adaptations. It is likely that the relative efficiency of
performance (by a thermodynamic or comparative criterion) varies with the particular
biological role. For a given individual, performance in each of the biological roles occupies
some position in the population frequency distribution. The relative net performance
across the biological roles determines fitness (*w*). A further difficulty to an already operationally challenging estimate of fitness is that substandard performance may not entail an
equivalent reduction in fitness across the different biological roles. The path follows the
general prescription of S. Arnold (1983), P. Wainwright and Reilly (1994), and Betz (2006).

species or assemblages. The next step is the most challenging: the specification of likely biological roles (B), some of which may constitute historical adaptations. From the organismal perspective, this status is irrelevant. What matters is solely that the B_n roles make a collective contribution to fitness (w). The estimation of w with respect to morphological variance can be viewed as a practical means by which functional morphology aligns with the scope of evolutionary morphology.

Operationalizing this approach presents diverging degrees of difficulty, depending on the variables being considered. A joint's articular geometry (J) is a feature that, with today's technology and morphometric techniques, can be described quantitatively as a continuous variable, rather than via an artificial meristic scale.[1] From this geometric information, a performance measure (P) can be estimated as a function of J. Variance in J represents some trade-off between the essential functions of mobility and stability. This measure of performance, by itself, is inadequate to map onto fitness in a meaningful way. The important linkage is that of the relation of P to biological roles (B) that are actually practiced, which may well be manifold. These roles will have aggregate effects for the fitness (w) of the character. Complicating matters further is that there are other morphological variables—here, the length (L) of the element in question—influencing the same or different sets of biological roles that also contribute to fitness. Add in the likely covariance between J and L, and it is an understatement to note that this framework, elementary as it is, still presents some difficulties. There is, however, a simple message embedded in this quest to link morphological characters to fitness: the critical consideration of how a morphological variable maps onto performance, and whether variance in performance has selective consequences *in the environment in which it is employed*.[2] We have astonishingly little data on performance variation under natural behaviors.

The enumeration of biological roles is not standard practice, so feasibility and formalization of identification need to be worked out. Comparative behavioral investigations will be decisive in determining whether such enumerations should even be attempted in the paleontological context. Phylogenetic bracketing of behaviors will also be indispensable, but this is certainly not foolproof (see below).

The secondary literature in paleoanthropology is remarkable, in that methodological decisions in functional morphology require no justification. Two renowned catalogs of primate fossils, *Evolutionary History of the Primates* (Szalay and Delson, 1979) and *The Primate Fossil Record* (Hartwig, 2002), advocate for the centrality of functional morphology in primate paleobiology, but they offer no overarching analytical principles to guide interpretation. Stringer and Andrews's

(2005) commentary on functional morphology is simply that it works, given that anatomy and behavior covary. Langdon (2005) provides a pithy discussion of the difficulties of adaptive inference, but the research practice is only discerned through example. A general sentiment of the field is that if you understand how the anatomy associated with a fossil performs and is used, the adaptive inference follows, without issue. The need for a theory of morphology is not discussed, because the question is almost never considered in the first place. Oxnard (1973a, 1973b, 1980, 1984) saw quantitative descriptions of morphology as the window into adaptive inference; he called these "mathematical dissections of anatomies." His interest was squarely on morphometric distance as an indicator of adaptive divergence. For him, shape,[3] numerically described, was transcendent, because it contained information about function, evolution, and development. Oxnard's objection to australopiths as human ancestors was, first, a functional argument, followed by a phylogenetic objection for the direct role of australopiths in later human evolution. His optimism that a numerical reduction of morphology clarifies the adaptive landscape remains highly influential, even though performance attributes will not align neatly with the patterning of multivariate morphospace.

The Great Escape Hatch: More Fossils Will Fix Everything

Paleoanthropology's reputation as half-baked paleobiology is often blamed on its inventory. The hope is that the interpretive problems plaguing the discipline will resolve once the number of fossils reaches some critical threshold. While this was a fair complaint 50 years ago, it is not terribly convincing anymore. If sample sizes are the real problem, then inferences should be expected to improve and theories to be honed as a simple consequence of larger samples. While this might appear unobjectionable in the abstract, in a concrete sense it is less defensible.

If one subscribes to the idea that there are species- and clade-level adaptations, then there should be sharp (i.e., minimally overlapping) boundaries between species with regard to morphological variables. This is more likely to be observed among small hypodigms. These boundaries will only become fuzzier as fossil samples get larger, and the ensuing taxonomic arguments will certainly erode the credibility of species-specific adaptations and behaviors. The irony is that "certainty" is as likely to decline with better samples as it is to improve. Order-of-magnitude improvements in fossil hominin inventory have not reduced disciplinary debate in any discernible way.

Does the General Approach Matter?

An unsettled question is whether inductive or deductive approaches to human evolution are more effective, in part because the distinction between them is not always clear in practice. Wolpoff (1980) argues for hypothetico-deductive falsification over induction, in the belief that falsification is more tangible than confirmation, especially in the paleontological context. Sober (2000) reasons that inductive inference suffers from the intrinsic problem of extrapolation from too few observations; i.e., generalizing from particulars is expected to fail sooner or later as sampling broadens. Paleoanthropology has no shortage of examples. Either approach, however, requires a set of assumptions through which to frame hypotheses. The amount of attention given to what these premises entail may be more consequential than the choice of approaches. Framing hypotheses in testable terms is de rigueur in functional morphology, and this sentiment is palpable in paleoanthropology today (cf. Daegling et al., 2013; Strait et al., 2013). Because there are lacunae in paleoanthropological data, however, the question arises whether falsifiability in the hominin fossil record is just a collective fiction, or whether a proscription of the practice is even helpful. In other words, are the difficulties of functional inference attributable to epistemology or methodology?

To answer that question, one must ask what the paradigm is that governs paleoanthropology. The requirement or sentiment that disciplinary focus should be on a unique set of species undermines explanatory power at the outset (Cartmill, 1990). Biological anthropology would then identify itself as something partially outside of, rather than encompassed by, biology in general. The long-abandoned single-species hypothesis envisioned culture as the primary human adaptation, and this defined the hominin niche. A sharp demarcation of culture either precludes or undermines comparative analyses involving anything other than hominins. Functional morphology in paleoanthropology is much better off if this is not true, but there is no denying the impact of culture on adaptation, by any definition. While this should have no effect on the biomechanical assessment of performance, with cultural adaptation, the possibility always looms that a behavioral workaround exists for any mechanical or ecological challenge. If the hominin niche is culture, then a workable paradigm with respect to functional inference has to resolve the rather obvious dilemma that cultural selection in *Homo* is qualitatively distinct from that in *Ardipithecus*.

The variability selection model of hominin adaptation (Potts 1996, 1998) can be seen as a refinement of the "culture-as-niche" idea that emphasizes behavioral plasticity in general as a hominin strategy. The name of the model is unfortunate, as—taken literally—it is an oxymoron.[4] Variability selection outlines a scenario in which species can evolve to function effectively in different environments without being particularly adept in any of them. Grove (2011a, 2011b, 2014) employed simulation algorithms to demonstrate the plausibility of the model under a variety of initial conditions and stochastic parameters. The operation of culture is implied in the model, although the form of Potts's (1996, 1998) model is certainly not restricted to cultural organisms. Instead, the guiding principle is that a competent eurytopic species can outlast stenotopic ones, given a degree of environmental instability.[5] Within the bounds of broader primate ecology, hominins are generally capable in many diverse circumstances. This suggests that for hominins, biological roles that are not (strictly speaking) adaptations may assume a greater role in determining fitness than they do in other clades.

The idea that paleoanthropology needs a paradigm distinct from paleontology implies that evolutionary biology, by itself, is inadequate to explain hominin emergence and diversification. It follows that viewing hominin origins as an outcome of historical contingency is also insufficient as an explanation; at the very least, it is unsatisfying. Nunn (2011) argues persuasively that the analytical problem Cartmill (1990) suggests—singular events defy explanation—is not intractable at all. In practice, and to some extent out of necessity, functional morphologists have been loath to embrace human singularity with respect to hominin evolution.

Despite a history of regarding human biology and evolution as unique, paleoanthropology (as a discipline) does not have a corresponding paradigm that neatly encapsulates the research programs active within it, other than evolutionary biology writ large. Claims of human uniqueness are recognizable and refuted through examples: tool use and menopause are but two instances, and the cognitive basis for language is not as exclusive as was once supposed (Seyfarth et al., 2005). Conflicting interpretations of the same data may provide evidence of paradigmatic tension within the field; i.e., structuralist versus functionalist perspectives on variation, multiregionalism versus out-of-Africa models of modern human emergence, or punctuated versus "gradualistic" modes of evolutionary change. To ascribe all disagreement to the tyranny of theoretical perspective, however, requires two very dubious assumptions: (1) all studies are of equivalent quality with respect to data and their analysis, and (2) the alternative methods applied to a particular research question are equally effective for answering it. Methodological sufficiency is a necessary consideration before declaring paradigmatic crisis.

Multiscalar Approaches to Functional Inference

One can artificially divide approaches to adaptive inference in paleoanthropology into a binary of top down and bottom up. Top-down approaches are usually argued on an ecological scale: "man the hunter" (R. Lee and DeVore, 1968), variability selection, and the turnover-pulse hypothesis (Vrba, 1985) are examples. In these models, adaptation is viewed as a species-level exercise in problem solving. The environment is the driver of evolutionary change, and morphological adaptation is assumed, if not manifest. *Why* certain traits represent adaptations may be explained in some depth, but *how* they arose in a morphogenetic sense is not explored. Consequently, structuralist ideas of constraints are absent, other than receiving a nod to acknowledge that they exist. A top-down model can be presented as an umbrella hypothesis, where there is a finite set of environmental variables that explain adaptation. Explicit measures of biomechanical or functional performance are not required; approximations suffice.

Any atomistic study of morphology that builds an inference of function from the details of anatomy represents a bottom-up approach. Chapters 4 and 5 are replete with examples. Links to environment and ecology may be drawn, but the source of inference is functional capacity or biomechanical performance. The species is less clearly monolithic in such research, because variation is impossible to ignore completely. Studies that utilize ontogenetic, morphogenetic, or mechanobiological data also qualify as bottom up. Ostensibly, bottom-up approaches are more process-focused with respect to the organism. These do not necessarily test for a fit between function and environmental variables but presume that they exist. Alternatively, as a practical measure, the environment may be reduced to one or more critical tasks to be competently performed.

With respect to functional inferences, top-down studies require only coarse considerations of morphology, while a bottom-up approach depends on a more precise analysis. The bottom-up approach is also more compatible with a structuralist outlook, although consideration of morphogenesis and constraint is rare in paleoanthropology (an early exception is Lovejoy et al., 1999). Functionalist perspectives dominate the literature on both top-down and bottom-up models. In reality, this dichotomy of approaches is forced, but I draw the distinction to emphasize that macroscale (i.e., climatic) studies, while attractive for identifying requirements of ecological adaptations, provide little guidance for understanding the emergence of certain traits, but not alternative ones. The question, What is bipedalism for? is within the scope of a top-down approach. The question,

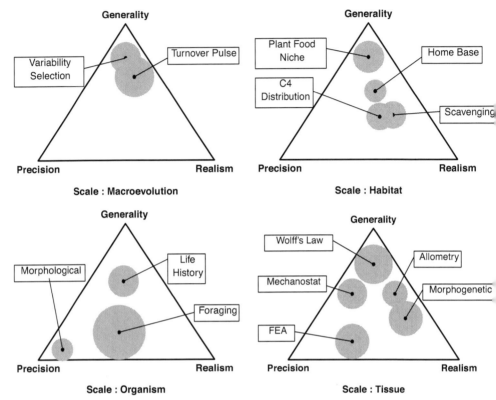

Figure 3.2. Models of behavioral inference in paleoanthropology. Sept's (2007) synopsis included modeling approaches at the macroevolutionary, habitat, and organismal levels; added here is a tissue level, with specific reference to skeletal material, including fossils. She noted that inferential models cannot achieve high precision, broad generality, and realism simultaneously. At evolutionary scales, models are very general, while the achievement of precision is largely determined by the nature of variables examined at other levels. At the organismal level, morphological studies can be very fine grained, but as illustrated here, precision should not be confused with realism. At all levels, realistic models are elusive in paleoanthropology. This is remedied to some extent by using spatiotemporally restricted ecological data and a variable choice that can be applied to paleontological contexts (e.g., taphonomic data applied to scavenging behavior; see Blumenschine, 1986). Mechanobiology is situated in the tissue-level domain and is productively incorporated into all of the models displayed here. The patterning I have produced is admittedly idiosyncratic, and there are potential objections to this scheme that could form the basis for useful disciplinary arguments. For example, the placement of morphogenetic models and finite element analyses (FEA) could be switched and justified, depending on one's opinion of the validity of the simplifying assumptions utilized in the respective approaches. In either case, it is far easier to increase precision rather than realism, but realistic models are more useful. Allometric studies, which often consist of a coarse assessment of function through elementary models, such as beam theory, have been influential precisely because they can be generalized across contexts and are empirically grounded. Based on Sept (2007).

How is bipedalism efficient? is more satisfactorily addressed in a bottom-up context.

Sept (2007) provided a heuristic for relating the scalar specificity of competing models of human evolution to the scope of their informational content (Figure 3.2). She envisioned competing utilities that models satisfy to various degrees: they can be precise, generalized, or realistic, but not all three simultaneously. Her original conception examined models operating at organismal, habitat, and macroevolutionary levels, while in Figure 3.2, I have added a tissue scale below that of organism. What is generally lacking at all levels are models that are realistic. The distinction between precise and realistic is clarified if we substitute "accurate" for realistic and think of the two concepts in mensurational terms. Precision means one can model details with good discrimination, whereas accuracy refers to a model that is in the vicinity of a correct prediction. Accurate models are realistic; precise ones may or may not be.

Mechanobiological models occupy the domain of the tissue scale, as do other approaches that model stress fields and tissue-level responses to environmental stimuli. Finite-element models have become commonplace in paleontological investigations (Rayfield, 2007), but they are not easily generalized. (Such models are recruited when generalized models are deemed inadequate for a research question.) Their level of precision is fully adequate, but accuracy is unknown for fossil material. Such models can be validated indirectly, however, through neontological data.

Mechanobiological models, properly formulated and operationalized, can target all three utilities. A model that simulates a specific load history to predict bone growth in terms of size and alterations in geometry is mechanically determinant and allows precision. An example is provided in Chapter 6. Alternatively, a model that contrasts various sets of stress stimuli to examine differences in cortical thickness would represent a generalized model that could inform studies of long bone robusticity (Pauwels, 1980). Finally, realistic models would have to be based on neontological experimental data at the outset. They would be unlikely to be immediately applicable to paleontological questions, but such models could subsequently be adapted to address them.

The Comparative Calculus

Functional inference has historically been dominated by comparative approaches. In both paleontological and neontological contexts, these methods are indispensable. While a biological role of a morphological feature can be directly observed in a living animal, and its mechanics and conformity to a prescription of performance specified and measured, for an understanding of fitness, comparison with other individuals is required. In the context of fossils, the need for comparative data is mandatory, since biological roles must be inferred, rather than observed. All functional inferences in paleoanthropology regress to comparative data.

The benefits of recruiting comparative data for paleobiology are transparent. It is far easier, and more convincing, to use comparative data to generate functional hypotheses than to evaluate morphology on its own terms and reconstruct alternative functions that are based only on performance criteria. Even if employing such an approach (discussed below in the "Paradigm Method" section), the morphologist eventually seeks out comparative data for validation. Part of the appeal of comparative approaches is that they are easily applied and are persuasive on a correlational basis (Nunn, 2011). A formulaic approach is common. If, in one or several living species, there is a trait X that is associated with a biological role Y, then the presence of X in an extinct taxon provides circumstantial evidence that Y is the biological role in the fossils. The strength of the association between X and Y basically determines the strength of the argument. If X is always linked with Y, and Y has a singular biological role, then Y fulfills that role in fossil species having trait X. The necessary assumptions of this logic are somewhat daunting, however. They presume that the range of possible biological roles for X have been circumscribed by the extant comparative survey. They also assume that the osseous features of X are a sufficient demonstration of equivalent performance with respect to Y. If the formula is to have any explanatory force, a one-to-one mapping of form to function not only must be possible, it must be likely.

The preceding also implies that the trait in question is easily and unambiguously defined. Discrete traits are preferred, because they are unambiguous, but, as a result, actual variation is often ignored. Unless the character states in question are a simple case of presence/absence, their "discrete" status is usually a fiction and, in the past, indicated a reluctance or inability to quantify size and shape. Modern morphometric techniques are fully capable of capturing the continuous variation that bridges previously assigned categorical character states. Since there is reason to expect that within the range of mechanical capacity, functional performance is gradational, treatment of the corresponding morphological variables as continuous is preferred.

The means of drawing functional inferences in comparative approaches is generally relational, rather than processual. The question of how traits came into being or the developmental basis for trait expression are easily sidestepped, while the association of trait states and biological roles is touted. Thus an ostensibly correct assessment of a biological role can be made without explicit reference to an evolutionary or ontogenetic process. As a result, comparative techniques are incomplete explanations.

The question also arises as to whether anatomical-functional associations in paleontological contexts effectively identify adaptations. The biological roles inferred, whether they are true historical adaptations or not, cannot be tested in the conventional sense. An association can be undermined, but not completely falsified, by counterexample. This frequently encountered tactic involves demonstrating an exception to an association. Attempts to evaluate Dart's (1925) inference that foramen magnum position could indicate bipedality have pursued this strategy (see Ahern, 2005; Russo and Kirk, 2013). Ohman (1986) argued that the near universality of a univertebral articulation of the first rib (i.e., to the T1 vertebral body, to the exclusion of that of C7) was functionally associated with bipedality. The association was based on claims that the more caudal articulation had consequences for thoracic shape, costovertebral mobility, and positioning of the inferior trunk of the brachial plexus. Stern and Jungers (1990) offered a rebuke involving a reevaluation of anatomical consequences of univertebral articulation, as well as a rebuttal to the claim that this was unique to humans. The latter was accomplished by sampling nonhuman primates and demonstrating that such articulations occurred with nontrivial frequency in hylobatids, indriids, and bamboo lemurs. They considered this comparative demonstration to be "more important" than their arguments, casting doubt on the functional consequences of univertebral articulation.

Rules of Engagement

References to the "comparative method" imply a monolithic, one-size-fits-all approach to functional inference. A diversity of practice belies this sentiment, although Kay and Cartmill (1977) offer procedural guidance for adaptive inference with respect to the fossil record. This prescription foreshadowed Cartmill's (1990) warnings with respect to an explanation of truly unique historical events.

In this scheme, adaptive inference is based on four criteria. First, any trait that is under examination for functional inference must be represented in the neontological context. Second, that trait, expressed in living organisms, must have some discernible singular function. Third, the trait in question must not have been fixed prior to acquisition of its current function. Finally, all trait attributes or features of multitrait character complexes must have a tangible connection to the current function. If these conditions are satisfied, then a reliable adaptive inference is possible.

Collectively, these stated criteria underscore the major difficulties of functional inference in paleoanthropology. The formalization is purposely restrictive and makes plain that the mapping of morphology onto adaptation must be direct if a confident assessment is to be made. While Kay and Cartmill (1977) recognize the value of a nonhistorical (i.e., utilitarian) concept of adaptation (C. F. Ross et al., 2002), their prescription is for recognition by the historical definition. Knowing the relative (i.e., phylogenetic) timing of trait appearance is the critical obstacle in this instance (Lauder, 1981).

This attempt to formalize comparative procedures is theoretically sound and methodologically helpful, but it does not solve all the deficiencies of comparative analysis. The insistence on one-to-one mapping of traits to functions enjoys no consistent empirical support (if it is investigated at all) and is essentially an appeal to parsimony. Parsimony is not inherently invalid, but in this case it is more a tool of first rather than last resort. The other problem left unsolved is that what amounts to a strictly correlational analysis of traits or trait complexes to inferred functions is seemingly explained by natural selection, although there is no independent corroboration that the requisite selective events have occurred. Fortunately, there are other approaches that can at least partially work around these problems (phylogenetic brackets and biomechanical reduction, each discussed below).

There is nothing inherently problematic about Kay and Cartmill's (1977) insistence that the object of inference is historical adaptation, in the sense of both Gould and Vrba (1982) and Lauder (1981). The third and fourth conditions (no fixation prior to current function, and all aspects of trait complexes relate to that current function) are necessary for the discernment of historical adaptation, but they do not govern the possibility of larger sets of biological roles. What this means, practically speaking, is that behavioral reconstruction is restricted if there is strict adherence to the idea that adaptation is the sole basis for inferring functional utility. The challenge is to operationalize hypotheses of biological roles in paleontological contexts. This also depends on comparative data sets; i.e., ones that provide information on multiple functions associated with "single" morphological traits.

Is Process Discoverable via Pattern?

The data utilized in comparative approaches are not unambiguous reflections of the evolutionary processes that underlie them (Leroi et al., 1994). The evolution of traits results, in part, from selection from the pool of phenotypic variation within a population; i.e., selection acts within a focal species. In evaluating functional morphology, however, the comparisons are ordinarily drawn between species. For example, a typical evaluation of australopith morphology relied on the degree to which one or more of the early hominin species differs from samples of chimpanzees, modern humans, and perhaps contemporaneous *Homo*. The contrasts are evaluated statistically from morphometric variables, which will reflect biomechanical functions with varying fidelity. Australopith performance for functional tasks is judged relative to other taxa, usually by a criterion of mean difference. It is not obvious, however, that such an assessment tells us anything decisive about the process of adaptation within the focal species.

If, in fact, the taxa being compared are not sympatric, then we should be even more skeptical about interspecific differences providing insight into the nature and degree of adaptation relative to ecological conditions (Figure 3.3). In the case of sympatry, a focal species potentially does have interspecific competition to contend with, and mean performance differences among species can have major consequences.

A few different things could happen: local extinction, emigration, character displacement from competitive exclusion, or perhaps a saltational increase in foraging efficiency, owing to a behavioral innovation. In each of these scenarios, the functional capacity of competing species can impact the evolution of the focal species, but the direction of morphological change is not governed exclusively by the altered selective regime. The mean difference in species performance is a reliable indicator of adaptive difference only if the performance variables in question are utilized in identical fashion.

While the role of sympatric species in conditioning a focal species' evolution and adaptation is tangible, in allopatric comparisons, more severe skepticism is called for (*contra* Williams, 1992). Mean differences in relative tooth size between Asian macaques and African colobines, for example, probably tell us more about phylogenetic and allometric effects on morphology than dietary and foraging adaptations peculiar to the individual species. There are regularities in primate morphology and behavior that transcend local ecological differences. Low intermembral indices are consistently associated with elevated leaping frequencies across a variety of habitats. The mechanical basis for this is well understood. The mistake arises in assuming that mean differences in this ratio are comparable energetically across all contexts or environments, or that a range of values is specific to a set biological role. To assume so is to equate statistical distance with adaptive divergence or performance attributes.

Similarly, disjunction in time dilutes the explanatory power of comparative inferences as much as disjunction in space. From the standpoint of understanding

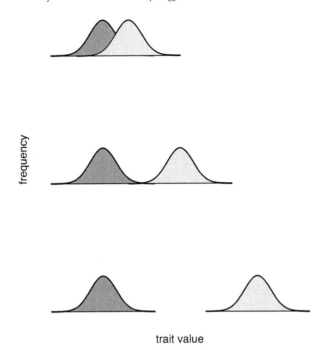

trait value

Figure 3.3. Character displacement is intelligible in sympatric contexts in which two taxa are in competition for resources (*top* and *middle panels*, showing initial contact and following sustained selection, respectively). Morphological divergence indicates shifts of function and will be interpreted as instances of adaptation. When there is complete separation of trait expression (*bottom panel*), the logic is extended without invoking displacement fueled by competition: distinctiveness in morphology is still indicative of altered function and, by extension, performance capacity with respect to resource exploitation. It is worth considering whether allopatric contexts should be interpreted in the same way for functional inference. Comparative skeletal investigations among primates interpret the differences in the bottom panel identically, whether or not the contrasted taxa are sympatric—and they usually are not. This is unproblematic if the performance capacity implied by morphological distance results in identical differences in resource extraction in the allopatric context. This would seem to require homogeneity of environments, which is an unreasonable proposition.

the adaptive significance of upright walking, it makes little sense to compare the bipedal efficiency of *Australopithecus afarensis* with that of modern *Homo*, since the former did not have to outperform the latter.[6] This yardstick of comparison answers the question of energetic efficiency, but biological role and adaptation are only tangentially informed by it, assuming the relevant energetic variables have been correctly identified and measured. A thermodynamic criterion of adaptation relies on an understanding of relative efficiency, but, with respect to natural selection, efficiency is always dependent on an environmental context in a particular time and place. If the context is completely unknown, then credible adaptive inference is generally not possible. An accounting of the extrinsic environment's role in organismal energetics would appear to provide the necessary information to detail metabolic costs of activity, but there are also intrinsic factors that must be considered. The most obvious of these is the effect of somatic size on metabolic rate, but short- and long-term physiological changes during the lifespan also play an important role. Pontzer (2017a) notes that the reality of metabolic adaptation—in which physiological changes can alter energetic efficiency—makes the accounting of energetic costs more nuanced than is generally appreciated. This means that increases in activity cannot necessarily be assumed to add energetic costs by a strictly linear criterion. One example would be the decrease in basal metabolic rate that follows sustained athletic training (Westerterp et al., 1992).

A neglected source of information in functional morphology, if not evolutionary biology in general, is intraspecific variance (Hallgrímsson and Hall, 2011). Given the default approach of mean differences between species being essential to functional inference, variance, more often than not, is thought to be the enemy of insight. It is seen as noise when sample means are imbued with special significance, whether the immediate goal is identification of biological role, adaptation, or biomechanical performance. At the very least, variation provides needed guidance in linking functional inferences with evolutionary process (Nunn, 2011). Specifically, it may be viewed as an index of evolutionary potential. Provided that constraint is relatively fixed over short time intervals, greater variation should allow a population to have more selective options (Van Valen, 1965). For a given functional trait, a population with a highly restricted variance has limited opportunities for change. Two populations can have the same mean trait values, and thus share similar functional/mechanical capabilities, but their evolutionary paths forward are quite different. Lack of variance can arise from divergent processes, including canalization, intense stabilizing selection, or fixation by drift. The assumption that population means represent a local optimum shaped by selection (with "local" here being shorthand for stochastic environmental variation) is reasonable for drift-immune (i.e., large) populations. Based on this assumption, scatter occurring around that mean can be thought of as noise; on the other hand, it is worth considering how quickly performance degrades away from that optimum, and whether it has meaningful fitness costs (Figure 3.4). An extreme example would be a uniform distribution. A mean trait value could be calculated,

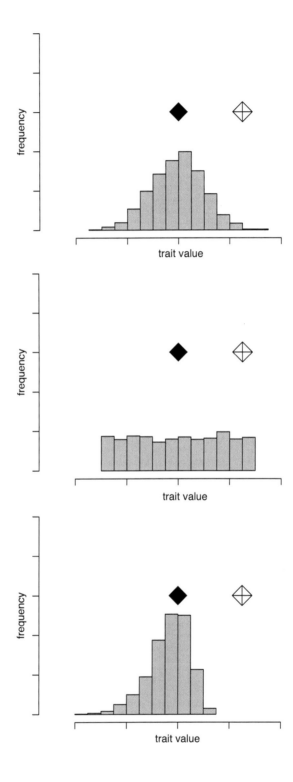

of course, but in this case, equivalent frequencies across the range would rightly invite skepticism about whether the mean tells us anything in terms of past selection or functional optima. In such a case, confidence in the biomechanical significance of the trait in question could still be high; i.e., an ideal performance condition could be specified quantitatively. An observed lack of coalescence around the optimum value, however, would strongly suggest that the variation in performance has an insignificant impact on fitness. Alternatively, it could indicate different biological roles across the species' range (the "broad niche" of Van Valen, 1965). Addressing these possibilities requires asking the relevant questions in terms of research designs that are focused on population-level variation. Such data are hard to come by in primate biology, and more so for Plio-Pleistocene hominin samples. Nonetheless, by ignoring patterns of variation as potentially informative, we miss an opportunity to construct specific hypotheses about the underlying evolutionary process producing morphological change.

Figure 3.4. The form of the population distribution of morphological traits informs the functional significance of trait values. In each graphic, the population distribution is shown as a histogram, with two observed trait values superimposed: one near the median (*solid*), and one displaced toward the right in each distribution (*hatched*). *Panel A* depicts a normal distribution. Under the assumption that a population is at an evolutionary equilibrium at the time of sampling, conventional wisdom states that the median value is close to maximum fitness. The value of the hatched trait is presumably suboptimal as an adapted state, but it may instead be exapted to particular environmental changes and could become rapidly selected, given its facilitation of a novel biological role. In *Panel B*, the population distribution is uniform, and there is no basis for thinking that the *solid* trait value is any more fit than the *hatched* trait value. The form of the distribution implies that the trait values have no meaningful impact on performance, because proximity to the median in this case has no predictable effect on fitness. The form of the distribution might also indicate the ability of individuals to use behavioral adjustments to achieve functional competence across the range of trait values, or it may reflect ecological variation and concomitant differences in biological roles across the population range. In *Panel C*, the asymmetry of the distribution suggests that the cost in fitness for being situated away from the mean also is asymmetric The *solid* trait value in this case may not be close to a performance maximum; the truncated *right tail* suggests a rapid decline in fitness at higher values, and a sound evolutionary strategy may be to avoid the *right tail* altogether developmentally (see Williams, 1992). The *hatched* trait value is obviously unobserved in the underlying population. Presumably, this indicates some change of function associated with a novel biological role, perhaps involving a quantum escape from modularity (see Young et al., 2010). These functional explanations are not the only valid ones for elucidating the observed distributions. The default assumption is that skeletal morphometric traits conform to a normal distribution; as these examples demonstrate, if this assumption is not true, then interpretations of adaptation can be productively revisited.

The Paradigm Method

Rudwick (1964) concisely argued for a formalization of a functional morphological method arising from from paleontological data. This "paradigm method" was inspired by the perceived need for testing—rather than blithely assuming—the default status of traits as adaptations implied by neo-Darwinian orthodoxy. He contended that reliance on comparative approaches amounted to pseudo-methods, in part because inferences from such approaches were constrained by the anatomical variants available in comparative material. Consequently, the full range of possible adaptations could not be appreciated in comparative contexts. Advocating for employment of analogies to machines (real or imagined) rather than organisms, he conceived of paradigms as models of "mechanical fitness." For a trait or trait complex, alternative paradigms could be articulated as optimal solutions for particular tasks, with the reconstructed functional capacity of the trait evaluated relative to paradigm performance. The function displaying the highest relative efficiency identified the correct paradigm, which revealed the likely adaptation. The approach even promised to shed insight on the quality of the adaptive trait: the greater the correspondence of the reconstructed function to the paradigm, the greater the degree of adaptation.

While the problem of interest for Rudwick was one of historical adaptation, it is nevertheless clear that what the paradigm method is exploring forms a broader utilitarian concept. Biological roles are tested just as easily and identically, given that "functional inference involves an analysis of adaptation only as a static phenomenon" (Rudwick 1964:38). He was also wholly unconcerned with teleological underpinnings. In his view, machine analogies were useful, because natural selection could be expected to produce machine-like organisms (*contra* R. Smith, 1982). Traits were identified as adaptations by treating them as engineering problems.

The paradigm method has been attractive to paleoanthropology for two reasons. First, it allows functional inferences to be drawn in the absence of environmental data. Taking this advantage a step beyond, it raises the possibility that performance-based inferences provide independent data for paleoecological reconstruction. Second, the method preemptively rejects Cartmill's (1990) axiom that unique historical events present insurmountable interpretive problems. If unique events are reframed as unique morphology (to which fossil data eventually are reduced), then one only needs to create a set of new paradigms to evaluate them. Thus the inability to infer an adaptation is only a failure of imagination.

Neither reason is sufficient to validate the paradigm method as a preferred approach for functional inference. As has been noted countless times, the invocation of a functional capacity to demonstrate presence of an environmental factor (e.g., arboreal traits prove the presence of trees) cannot be subsequently used as validation of the adaptive inference. Otherwise, a misplaced optimism emerges, where paleoecology can simply be thought of as an extension of morphology. The paradigm method is also optimistic to a fault with respect to its ability to understand unique morphology. There is always the possibility that the correct paradigm is never articulated in a given case, and, at the same time, a reliable functional reconstruction will always be in doubt if homologies are ambiguous and no clear analogies exist. Lauder (1995) remarks that because the paradigm is informed a priori by the morphology in question, the method cannot provide an independent assessment of function.

Rudwick (1964) appreciated these problems and saw the paradigm method as an imperfect, probabilistic approach for evaluating competing functional hypotheses. Nevertheless, it is ironic that in an effort to accommodate adaptive inference within paleontology, the method recruits nothing from evolutionary theory, other than a general belief that natural selection is a near-ubiquitous, highly efficient phenomenon. Nowhere is there a hint of structuralist sensibilities, except for a caution that the mechanical properties of biological materials should be appreciated. Signor (1982) complains that the method opted unrealistically for a narrow, singular optimum, driven by selection, so it cannot deal with the prospect of multiple adaptive solutions. This criticism implies that Rudwick (1964) was indifferent to alternative adaptive or nonadaptive possibilities, but this is false: the paradigm method operates from the premise that nonadaptive morphology is immune to tests. The paradigm is heuristic and thus is not a formal model of adaptive evolution.

There are two additional major limitations to this method that are also encountered in other approaches. The first is that it does not confront the problem of trait origins in phylogeny or ontogeny (Rudwick deemed this "irrelevant"). The second is that the machine analogy ensures an essentialist outlook with respect to organismal function: variation is only thought of as a manufacturing quality-control issue. Given that the yardstick of optimal design is central to the approach, the paradigm method is seldom applied in isolation (Anderson et al., 2012; Grine and Daegling, 2017). It does, however, offer one tool of value for post-Gouldian assessments of adaptations: the absence of fit between morphology and a sensibly conceived paradigm provides some guidance about the operation of constraint (Gans, 1985, 1988; C. F. Ross et al., 2002), even if the specifics are as yet unidentified.

Analogy

Because the environment is effectively ignored, the paradigm method is an example of a very limited bottom-up approach to functional inference; it places a plurality of hope on morphology, with the remainder split between imagination and calculation. For this reason, it cannot be expected to yield much perspective on hominin paleoecology. For more comprehensive assessments of hominin paleobiology, paleoanthropologists often utilize analogies to living animals, in which one or more aspects of species ecology are thought to mirror those of fossil hominins. In essence, extant species are used as models for early hominin behavior. In one sense, since comparative and paradigm methods seek to infer behavior over a restricted scope of activity, the use of analogy can be thought of as a type of holistic or extrapolated functional morphology to the whole-species level. This characterization belies the very different means by which these approaches come to understand morphology. Comparative and paradigm methods conceptualize function as a matter of a finite physical capacity and its effect on performance. Behavior must, at a minimum, be compatible with that performance. Function is defined biomechanically; i.e., quantitatively, with reference to specific variables. For analogies, behavior is known at the outset—it is the rationale for choosing the model in the first place. Behavior does not have to be commensurate with the morphology of organisms; in effect, there is no morphological problem to be solved. A biomechanical proof of competence is unnecessary, and the significance of morphological details will be ad hoc.

Analogies drawn to model early hominin behavior include wolves (Arcadi, 2006; C. Walker and Churchill, 2014), other social carnivores (Schaller and Lowther, 1969), baboons (DeVore and Washburn, 1963; Jolly, 1970; Swedell and Plummer, 2019), chimpanzees (Stanford, 1996), and bonobos (Zihlman et al., 1978). Entire volumes have been devoted to hominin ecological and behavioral reconstruction using analogies of nonprimate social carnivores (R. Hall and Sharp, 1978) and nonhuman primates (Kinzey, 1987). These works make it clear that there is no requirement for morphology to be congruent in detail or pattern between hominin fossils and the model species standing in for them. This is obviously problematic, because the strength of the narrative has an outsized influence on perception of an analogy's plausibility. There are good reasons to doubt that morphological congruence is expected on the basis of a different phylogenetic history alone. What matters is whether the morphology can be retrofitted to a suite of behaviors, which are often nebulous (e.g., hunting and gathering can be accomplished in myriad ways). Here there is far too much wiggle room from the morphologist's perspective.

As paleoecological models, analogies are insufficient to explain the evolution of morphology in processual terms, such as morphogenesis. "Beanbag" genetics for a time provided an out, since the acquisition of bipedality, for example, was explained simply as a relentless selection of alternative alleles.[7] This hardly qualifies as a good explanation, as it can be applied with equal success to any historical change in anatomy. In any case, today there is no longer discussion of alleles *for* bipedality. The problem intrinsic to all analogies for hominin paleobiology is that they are expected to fail at some level of analysis; baboons are not bipedal, hyenas do not use tools, and wolves are not encephalized.

This last criticism is a source of annoyance for the authors cited above. The goal of analogy is to explain constellations of features as historical adaptations, but these features need not resemble those of hominins. Instead, they should be associated with solving a common set of ecological problems by different means. In this way, the value of analogs for hominin behavior is in the realm of hypothesis generation, rather than hypothesis testing.

As is the case for most scientific endeavors, the variance in study quality and utility is disappointingly large with respect to analogies for hominin emergence and evolution. Their value can be retrospectively evaluated by their influence on contemporary debates, as well as their initial impact. Arguably the most important analogy ever drawn to explain hominin origins was Jolly's (1970) invocation of *Theropithecus* as an ecological analog to early hominins. The specificity of the analogy provided tangible circumstantial data in support of the "seed-eating hypothesis": *Homo* is to *Pan* as *Theropithecus* is to *Papio/Mandrillus*. An exceptional aspect of that paper was marshaling specific features of skeletal and dental anatomy and linking these to ecological variables (with varying degrees of success). Jolly conceived of this as an exercise in the identification of parallelisms, with *Pan* and *Papio* serving as foils to expose spurious associations. Open-country foraging requiring manual dexterity and orthogrady in *Theropithecus* was used to argue for the same behaviors in the earliest australopiths (and discovery of *Australopithecus afarensis* was still a few years off).

As must be the case for any allospecies analogy, the adaptive meaning of the comparisons was sometimes forced, but Jolly's (1970) argument was the beginning of the end for "man the hunter," at least as far as hominin origins was concerned.

Regardless of whether seed-eating was the agent of hominin emergence, Jolly's hypothesis facilitated alternative origin scenarios. It also provided a relatively unromanticized concept of hominin divergence in the wake of Dart's (1957) osteodontokeratic musings. The lasting influence of the analogy was not so much that it solved the ecological enigma of hominin origins (its presupposition of a savannah habitat assured that it did not), but that it forced a shift in perspective about hominin uniqueness. The seed-eating hypothesis suggested that hominins were not *sui generis* but could, in fact, be considered just another primate and be comprehensible as such. The historical outcome is that much more comparative primate research followed, and it was deemed relevant for understanding hominin paleobiology.

Not all hominin analogies resemble umbrella hypotheses or general ecological models. Biomechanical analogies (e.g., DuBrul, 1977; Hutchinson and Gatesy, 2001) are common and take advantage of evolutionary convergence for developing comparative models. Although more limited in their explanatory scope, these models are more amenable to test.

Phylogenetic Brackets

An analogy for early hominin ecology based on chimpanzees, as opposed to theropod dinosaurs, would be regarded by most anthropologists as a more prudent comparison. This is not only because chimpanzees look more like people than *Allosaurus*, but also for the more explicit reason that humans and chimpanzees have a larger proportion of their evolutionary history in common. Homologies will presumably be easier to identify, certain behaviors are likely to be shared, many constraints will be held in common, and the detection of historical adaptations will be more reliable. With the advent of formalized cladistic methodology, the evolution of traits can be mapped onto phylogenies, and the distribution of traits on these trees provides an explicit topology of the distributions of traits among taxa. As a result, functional inferences and the diagnosis of adaptations should be less prone to ad hoc and anecdotal inference.

Witmer (1995) advocated employing the *extant phylogenetic bracket* as a means to infer soft tissue morphology in fossils and, in turn, permit a more direct and reliable means of functional inference. It relies on the use of outgroups and parsimony to guide decisions about soft tissue reconstruction, in the same manner as Lauder (1981, 1982) developed for discerning historical adaptation among hard tissue variables. The logic is straightforward, even if ambiguity is certain to arise in many analyses. Susman (1998) and Susman and de Ruiter (2004), for example, adopted a bracketing approach for reconstructing hand function in early hominins (Figure 3.5).

It is not too much of an overstatement to suggest that the use of the phylogenetic bracket is the primary methodological break between evolutionary morphology and functional morphology. In the absence of a temporally complete fossil record that characterizes rates of morphological change in an evolving population,

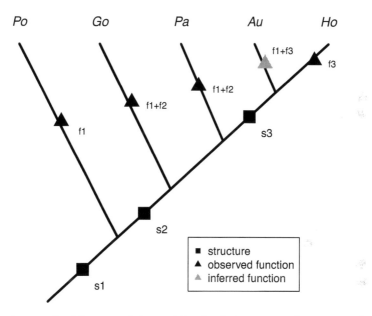

Figure 3.5. An example of the extant phylogenetic bracket for interpreting character transformations and function in the fossil record. Great Apes (Po = *Pongo*, Go = *Gorilla*, Pa = *Pan*) and humans (Ho) provide the bracket for interpreting the morphology of *Australopithecus afarensis* (Au). *Squares* represent reconstructed instances of character transformations (S), and *triangles* indicate associated functional capacities (f). S1 is a character complex used in climbing (f1); because of the distribution of S1 among living apes and the retention of some of its characters in the fossils, it is not considered lost until the final split of *Homo* from *Australopithecus* (see Chapter 4 for a contrary interpretation). S2 represents a character complex to facilitate knuckle-walking (f2). Parsimony favors S2 arising once and not in parallel. S3 represents the appearance of characters related to bipedality. This approach is useful for constraining speculative scenarios of character evolution. It is not foolproof, however, for at least two reasons. First, the phylogeny must be correct. Second, it depends critically on the included taxa. If *Ardipithecus* is inserted at the appearance of S3 in this example, the original interpretation is necessarily upended: extant ape-like climbing capacity (f1) is now disassociated with the morphological changes at S3. Chapter 4 also underscores that agreement on phylogeny and character states does not need to yield a consensus as to how S relates to f in each case. Modified from Susman and de Ruiter (2004).

the phylogenetic bracket provides, at least theoretically, a means of documenting trait evolution while permitting the diagnosis of homoplasy and identification of evolutionary processes (Nunn, 2011).

While the value of the phylogenetic bracket is transparent, even its advocates agree that the technique does not overcome the necessary but eventually inadequate assumption of direct mapping of structure onto function. This constrains outcomes in undesirable ways, in that by using extant taxa to infer extinct functions, it does not permit the detection of novel or unique functions from the fossil record. The hope of the morphologist in recruiting the phylogenetic bracket is that the comparisons drawn will unambiguously reveal evolutionary mechanisms. Leroi et al. (1994) argue that analysis of traits in terms of their phylogenetic distribution is insufficient to infer historical adaptation. The basis for this deflating conclusion is that different genetic mechanisms (pleiotropy, epigenetic interaction, correlated selection) can produce the same trait distributions in a given phylogeny. In addition, the phylogeny will treat traits as monolithic for each taxonomic unit, thereby ignoring the variations within taxa that fuel natural selection.

The phylogenetic bracket thus mitigates, but does not completely solve, the important problem of moving beyond morphological patterns to the discovery of the processes that produced them, but there are still sound reasons for utilizing such an approach. A phylogenetic perspective provides information on the evolution of traits that is nominally testable, even if a cladistic analysis cannot resolve the status of traits as adaptations. The pervasiveness of homoplasy, inferred through cladistic analysis—despite serving as an annoyance to the systematist— is an essential question for functional morphology (Wake, 1991). Homoplasies may represent adaptations or be evidence for the operation of constraint. In either case, they provide information about the bounding of morphospace, and these discontinuities deserve attention, as they are apparent possibilities that are actually impossible. An analogy might be Sewall Wright's adaptive landscape being figured in morphometric instead of genetic terms.

Biomechanical Reduction

In the paleontological literature, qualitative descriptions of function have incrementally given way to more quantitative biomechanical assessments. It is arguable that this has improved insight into adaptation, but it is certain that the characterization of function and performance has become more objective. Whether using a recognizable analog or not, a biomechanical model essentially provides a description of the physical behavior of an anatomical system. Long bones as beams

(Ruff, 2000), crania as shells (Demes 1985), and mandibles as levers (Greaves, 1978) are all reductionist in the sense that the morphology in question is modeled simplistically and subsequently evaluated in terms of structural integrity, efficiency, or performance, usually in terms of a single or a handful of variables. Consequently, any biomechanical appraisal provides an incomplete accounting of morphology, even as a description of function, since a biological role need not be known to conduct a cogent analysis. The question naturally arises as to whether numerical biomechanical descriptions are reductionist in the worst sense; i.e., the biology is oversimplified and the function that is modeled is isolated from all other organismal functions. This objection is another referendum on atomistic approaches to functional morphology.

The critique has force if a demonstration of biomechanical performance is immediately and unequivocally given the status of a historical adaptation. Unqualified claims of this kind are relatively rare, and, in such cases, the claim is as likely as not to be an accidental one, involving careless articulation. If, however, a biomechanical analysis is undertaken merely as a means of evaluating a functional capacity (e.g., speed, stiffness, force production), then the question of adaptation is a separate one, informed but not decided by the data on relative performance. What a biomechanical model can provide is a metric of potential adaptive utility.

Biomechanical analysis is thus merely one component of adaptive inference. It is difficult to conceive of it in any other way, given that adaptations cannot be identified outside of behavioral or ecological contexts, which are easily ignored in a mechanical study. The utility of biomechanical models in the evolutionary context depends on the general validity of the thermodynamic criterion for adaptation *sensu lato*. On balance, morphological configurations that save energy in terms of somatic maintenance have positive fitness effects. This is a necessary but insufficient condition, however, for identifying adaptation. Given the prevalence of natural selection, any anatomical complex that has evolved is mechanically efficient, biologically speaking. This can lead to the vacuous postulate that every fossil found represents an adaptation of some sort, historical or otherwise.

This premise can be addressed and overturned. For example, pleiotropy may act to maintain morphological features that have net deleterious effects (Leroi et al., 1994). From an anatomical (as opposed to a genetic) perspective, this could be evaluated in two ways. The first would be to examine covariance patterns and determine those that have negative effects on fecundity. This requires counting offspring and recording morphology over several generations—feasible for *Drosophila* in the lab, operationally prohibitive for natural primate populations, and practically impossible for fossil hominin assemblages. The second option is more palatable methodologically: examine covariance patterns from the perspective of biomechanical function to identify how the covariance may negatively influence performance.

The yardstick in this case is relative—not absolute—efficiency. The degree to which an anatomical complex converts muscular force to acceleration is far less relevant than how a population's morphological variation conditions the variance in performance vis-à-vis an actual or hypothesized biological role. If biological roles are malleable with respect to morphology, then a congruence of physical optima (i.e., what the biomechanical model is positioned to evaluate) to ecological optima is not a foregone conclusion. Comparisons of relative physical efficiency are not informative if the contrasts are inappropriate.

Calculation of relative efficiency has the apparent advantage that every parameter determining a functional capacity does not need to be known in absolute terms. Biomechanical models become analytically cumbersome as desired accuracy increases. If a point estimate of a mechanical property or performance variable is necessary, then every parameter involved in its calculation must be estimated. If one only has to know whether a given anatomical configuration or feature performs better or worse than another, then this requirement can be relaxed, to varying degrees.

The determination of relative efficiency can be formalized in terms of "phenotype sets" (Reeve and Sherman, 1993) if quasi-discrete expressions of a biomechanical variable or complex of variables can be specified. In this approach, an adaptation is deemed to exist if the extant (or most common) phenotype can be shown to be the most efficient one among the alternatives, real or imagined. If the context of the biological role is ignored, this reduces to the paradigm method.

Models can be charitably described as incomplete representations of reality, yet they are indispensable to scientific practice if the goal is to learn something. When presenting a model of human evolution, one has to expect that a justification of the model will be required.[8] There are several strategies for mounting a defense, some less cogent than others. You can declare the model to be unassailable, on the grounds that no evidence exists to refute it. Alternatively, you can declare victory by default, deciding that since other models are wrong or incomplete, one that has yet to be so labeled is therefore correct (see Szalay, 1975; Cachel, 1975; Jablonski and Chaplin, 1993). Neither strategy advances an argument productively. The preferred avenue of model defense is to enumerate one's simplifying assumptions (i.e., where one cut corners on realism) and evaluate the analytical consequences of those decisions. This is much easier to do with an explicit biomechanical model than it is with a more amorphous ecological analogy. Just as in an ecological model, simplifying assumptions are intrinsic to any biomechanical model. This sometimes goes unrealized, because in a biomechanical analysis, one can get to a fixed, singular answer that wrongly implies accuracy. Acknowledging the analytical shortcuts is the path toward model improvement. The objective of this type of defense is not to discover if your model is wrong (as it most assuredly is in some details), but to know where its deficiencies lie. Generally, a model with a more reasonable set of simplifying assumptions is a better one. The rationale for doing sensitivity tests in finite element models or evolutionary simulations is not to find a correct iteration (this is unknowable), but to see how sensitive a model is to an estimated parameter manipulation.

Morphogenesis through Mechanobiology

A growing bone is as subject to natural selection as a mature one. The proximate means of evolution of skeletal morphology is the alteration of bone development. These two statements justify the use of mechanobiological models for examining bone adaptation. They cannot be tested via paleontological data, but what such models provide is an opportunity to examine patterns of variation emerging from a finite set of morphogenetic rules. R. B. Martin et al. (2015) and Carter and Beaupré (2007) described a number of mechanobiological models in which rules of bone apposition and resorption are tied to states of stress and/or strain. Some of these models, based on relatively simple mathematical functions, succeeded in re-creating variations in ossification patterns observed in the skeletons of humans and other model species.

Skepticism about the utility of mechanobiological models for paleoanthropology is understandable, given that validation from fossils is out of the question. There are too few nonadults in the hominin inventory to provide an adequate sample for modeling the nuances of growth and development of particular skeletal elements. There are, however, important questions about variation in fossils that can be answered by such an approach but cannot be resolved via other methods, including whether modest species- or population-level differences in osteogenic stimuli can produce distinct morphology in the absence of large behavioral differences.

One such application is the adoption of agriculture, which was a revolutionary event in later human evolution. It is difficult to overstate its significance. The gracility of the modern human skeleton has been linked to instances of this event. Studies of a subsistence transition from foraging to agriculture have observed reductions in skeletal robusticity (Ruff et al., 1984; Larsen, 1995, 1997). While the possibilities for the decline in relative bone strength are manifold, a preferred hypothesis is that with agriculture, a qualitative change in labor occurred (Larsen, 2006). Basically, the idea is that agriculture created a general reduction in the daily stress stimulus on the skeleton, via a diminution in strenuous and/or prolonged activity. Whether this is intuitive is arguable, but there are a number of counterexamples that invite skepticism (Bridges, 1989, 1991; Peterson, 2002). When mechanobiological principles are applied to the problem, it becomes clear that a qualitative accounting of behavioral changes are inadequate to resolve the question.

The assumption in the bioarchaeological context is that general patterns of behavior while hunting and gathering are distinct from those involved in horticulture and agriculture. In the language of skeletal biomechanics, the daily stress stimuli in these populations are sufficiently disparate that the mean modeling responses should be different between them. Current orthodoxy suggests that the interaction of load frequency and magnitude provide this daily stimulus. Since intervals of inactivity are crucial for shaping the degree of mechanotransduction in bone tissue, the daily stress stimulus can be amended into the concept of an effective strain stimulus (Beaupré et al., 1990; Koniecynski et al., 1998). The effective stimulus is the subset of experienced loads that are actually transduced by the osteocyte network, which is the critical variable for assessing the potential of the modeling response. Load frequency, strain magnitude, and the number of loading cycles determine a daily stress stimulus. The difference between the daily stimulus and the effective stress stimulus is dependent on the number and duration of rest periods between loading bouts, because the osteocyte network quickly becomes desensitized during a loading bout, and there is a refractory period determined by the duration of inactivity. Turner and Robling (2003) developed functions, based on experimental data, that describe these effects.

Figure 3.6 provides a single contrast in activity patterns between two populations. In this example, the only difference in the stress stimulus is the number and duration of periods of rest (the total amount of rest over the day is identical). The patterning of this rest turns out to be highly consequential: depending on the details of the daily stress stimulus, the effective stress stimulus is either higher or lower in the population with fewer rest intervals. Running 10,000 simulations, in which the only difference between the populations is in the partitioning of the daily rest, produces a distribution in the ratio of the effective stress stimulus that is highly asymmetric (Figure 3.7). In particular, the population with more shorter-

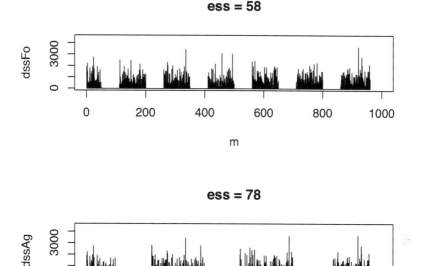

Figure 3.6. A daily stress, or strain, stimulus (DSS) does not translate directly into bone modeling, due to the fact that osteocytes are quickly desensitized to loads and have long refractory periods. The more appropriate index of osteogenic signals is the effective strain stimulus (ESS), which accounts for these temporally and frequency-dependent variables. In this example, an identical load history characterizes agricultural (dssAg) versus foraging (dssFo) activity over the minutes (m) of a 16-hour activity profile. The only difference is the distribution of rest periods (totaling 300 minutes in each case) over the course of the day, shown as periods of zero strain stimulus (the distribution of rest periods between the groups is arbitrary). The DSS for this example differs by less than 0.1% for the two groups and is entirely attributable to the fact that rest times are not coincident. The ESS for foraging activity is 58, and that for the agricultural activity is 78. This indicates a more pronounced metabolic response in the agricultural group. The form of the loading profile has a very profound influence on the relative ESS. The DSS profile was generated from random Weibull distributions with parameters 1.5 and 500 for strain magnitude, and parameters 8 and 2 for load frequency. In addition to the simulation shown here, an additional 9,999 were run to evaluate variability in outcomes (see Figure 3.7).

bout rest intervals has highly variable effective stress stimuli, sometimes exceeding those of the fewer longer-bout rest intervals by tenfold or more. Conversely, the population with fewer long-term rest intervals—in the cases when it has the higher effective stimulus—usually exceeds its counterpart by less than 50%. What is also noteworthy is that the effective stress stimuli are rarely equal or nearly equal

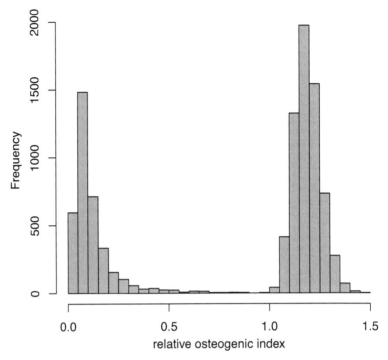

Figure 3.7. The ratio of the agricultural effective strain stimulus (fewer rest periods) over the foraging effective strain stimulus (more rest periods), calculated over 10,000 simulations, is reported here as a relative osteogenic index (OI) of consequent bone modeling activity. In most cases, the OI from fewer daily rest periods is marginally higher (> 1) than the OI under more rest periods, but in the rarer cases in which the OI is higher under more rest periods (i.e., foraging > agriculture), it is modally an order of magnitude higher. The implication is that very subtle differences in temporal loading profiles can have a quite substantial impact on subsequent osteogenic activity.

between the two populations. It does not matter which pattern we believe matches a foraging behavioral repertoire and which one reflects agricultural activity, it is unclear which skeletons should be more robust. The mean osteogenic index across simulations is 0.78, but the distribution of index values indicates that those in the vicinity of the mean are almost never encountered.

The lesson from this exercise is that the details of activity matter tremendously—not just in terms of the severity of loads, but also with regard to the patterning of inactivity. What is clearly needed are hypotheses beyond those specifying more activity, Specific behavioral data on foraging and agricultural labor are required, and a few moments of reflection are sufficient to convince one that neither will be homogeneous across populations and landscapes. Relying on skeletal robusticity to explain behavioral details in the absence of ancillary data is perilous.

The preceding example demonstrates the value of experimentally derived data for behavioral inferences. Direct investigation of physiology in the fossil record is obviously impossible, yet experimental data are invaluable for framing and testing functional and adaptive hypotheses in the paleontological context. Such data have also justified uniformitarian assumptions in the context of bone biology. The general applicability of dynamic strain similarity (Rubin and Lanyon, 1984a) and safety factors in bone (Biewener, 1993) make the modeling of long bone robusticity in the fossil record relatively straightforward. Similarly, experimental characterization of mandibular bone strain (Hylander, 1979, 1984) and masticatory loading (C. F. Ross et al., 2007, 2016) provided the interpretive framework for masticatory function in the primate fossil record (e.g., Hylander, 1988; Daegling and Grine, 1991; Ravosa, 1996, 1999, 2000; Daegling et al., 2014).

When descriptions of fossils lead to scenarios of hominin paleobiology, the persuasiveness of the argument, in the absence of neontological empirical support, becomes a matter of the skill of the narrator. A seemingly dissimilar issue, but one that is also resolvable with an appeal to experimental data, is that phenotypic plasticity is real and has proximate effects on fitness, even if its operation is invisible in the fossil record. It is worth asking whether our taxonomic schemes lead us into essentialist interpretations; experimentally induced plasticity suggests that this is a real danger (Ravosa et al., 2016). The materials of paleoanthropology are largely resistant to experimental manipulation, but it is certain that tethering functional-mechanical inferences to neontological experimental data improves our ability to evaluate competing hypotheses in the fossil record.

The Law of the Hammer

At any given time, a handful of analytical methods have had an outsized influence on interpretation of function and adaptation in paleoanthropology. In part this is because we are restricted to certain avenues of inquiry for uncovering behavior. Experimental manipulations *in vivo* are impossible, genomic variation is inaccessible, and the environment can only be reconstructed through proxy variables. These limitations are intrinsic to paleontology. We should be cognizant of the possibility that the applicability of certain methods to the fossil record is not necessarily suitable for all the questions we are interested in.

The law of the hammer stipulates that if you provide a student with this tool, under instructions that "here is your method of inference," the student will logically conclude that data are to be had by pounding on things (Maslow, 1966). In a world full of nails, it is an ideal analytical instrument. If, on the other hand, the most relevant variables are to be investigated by combustion, whether that hammer can or should be applied to their measurement is an important question.

Technological advances have spurred new avenues of research that were unimagined 50 years ago. Methodologies using isotopic signatures in tooth enamel to model hominin landscape use (Copeland et al., 2011) or sampling Neanderthal DNA to determine genetic introgressions into modern humans (R. Green et al., 2010) were not generally anticipated. Even so, many intractable issues of paleobiology identified in the 20th century persist. Has evolutionary theory changed to the same degree as our technological capacity? This is an apples and oranges question—impossible to answer intelligibly without a common criterion of change—but it raises the issue of whether historical shifts in research foci have been driven by changes in theoretical emphasis or instead are conditioned by methodological and analytical innovation.

Certainly neo-Darwinian synthesis influenced the way paleontologists thought about evolution, but in terms of behavioral and functional inferences, there was no innovative shift as to how morphology was investigated. Similarly, Gould and Eldredge's (1977) model of punctuated equilibria did not really prescribe a new methodology; it only admonished the morphologist to revisit assumptions about how temporal changes should be interpreted. Phylogenetic systematics (cladistics), in contrast, has arguably been more influential, since its specificity regarding species relationships allowed a consistent criterion for inferences of historical adaptation from morphology. In this case as well, however, the procedures for drawing functional inferences at the anatomical or organismal level was largely unaffected. Mechanics (articulated numerically or informally) has always served as the primary means of evaluating functional hypotheses.

Mechanics is not an analytical tool that, in isolation, answers behavioral or evolutionary questions. How to apply that performance information to these questions has been a persistent challenge for paleontology in general (Lauder, 2003). In the first half of the 20th century, functional inferences in the hominin fossil record were largely idiosyncratic. This was partly due to the paucity of the record and the inaccessibility of specimens. Morphology was scrutinized for taxonomic and phylogenetic assessments, and once those issues were settled, functional and adaptive hypotheses were then explored. The assumption was always that taxonomic distinction corresponded to a similar degree of adaptive or behavioral differences, without a perceived need to make these explicit (Montagu, 1955).

In the latter half of the 20th century, more-formal approaches to functional analysis and the inference of adaptation emerged, some of which have served—for better or worse—as methodological hammers in functional morphology. Allometry (discussed in Chapter 2) has proven to be indispensable; today, no functional analysis is taken seriously without consideration of potential impacts of size and scale on morphology. Increases in computational power over the past half century have given rise to other hammers.

Identifying covariation in morphological datasets is recognized as essential for understanding the evolution of traits and trait complexes in comparative contexts. Multivariate studies of morphology are ubiquitous today. As most of us are incapable of conceptualizing n-dimensional morphospace when $n > 4$, we utilize graphic procedures for summarizing how variance is partitioned via ordination (Figure 2.3). Such studies are valuable for describing how well taxa (or sexes) can be distinguished and which variables are effective in doing so. These techniques have also been championed as being suited for functional hypotheses (e.g., Oxnard, 1973a, 1973b, 1975, 1980). The loadings of variables on principal component (or principal coordinate, or canonical variate) axes do provide information on which measurements are influential in sorting specimens. As noted in Chapter 2, there is no reason to assume, however, that these variables possess any overarching functional or adaptive significance. They are simply effective discriminators of a particular set of taxonomic entities.

This does not mean that the multivariate hammer is inconsequential with regard to functional morphology. Multivariate techniques, including geometric morphometrics, can provide important insights into "evolvability"; i.e., the relative ease with which a given morphological trait or complex can be transformed (Wagner and Altenberg, 1996). This concept is particularly germane to functional morphology (as it pertains to paleoanthropology), because we want to know whether certain transformational paths are more likely than others, from a morphogenetic perspective (Wagner, 2000). Central to this endeavor is the concept of modularity—collections of traits that function and develop as a unit—which is inferred through examination of covariance among traits. Modules presumably influence the ability of natural selection to tinker with performance.

A more direct hammer for functional inference in paleoanthropology is finite element analysis (FEA). This is a theoretical mathematical technique that can provide a full-field depiction of stress and strain in anatomical models, including fossils. The advantage of FEA over other biomechanical modeling techniques is that it requires fewer coarse-grained simplifying assumptions about specimen geometry,

homogeneity, and material properties.[9] While the underlying principles and operationalization of this method has been known since the 1940s, limitations of computational power precluded its widespread application until several decades later,
with initial modeling of bones beginning in the 1970s (Huiskes and Chao, 1983).

Basically, FEA breaks down a prohibitively complex stress analysis into a large
array of manageable ones (the "elements" in Figure 3.8) and, in so doing, provides

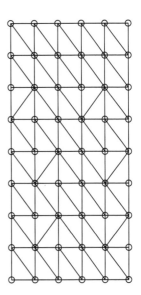

Figure 3.8. The difference between application of beam theory (*left*) and the finite element
(FE) method (*right*) in an idealized cross section. The former approach is easily applied to
large comparative samples, while the latter is ordinarily applied to one or a handful of
specimens and then generalized to populations. Beam theory requires a number of simplifying assumptions, including a high span:depth ratio and bending that occurs along planes of
symmetry, as well as material isotropy and homogeneity. Its primary advantage is relative
simplicity in the calculation of the stress field: the highest stresses are found farthest from
the section's neutral axis, where the normal stresses are zero. Higher stresses are indicated
by increasing grayscale; the relationship of stress to distance from the neutral axis is linear,
but discretizing the section (as shown here) is a suitable approximation in broad comparative contexts. FE models can describe the stress field under a variety of load cases, material
properties, and boundary conditions. Consequently, they are more versatile, albeit at the
expense of computational intensity. Stresses are determined in each element from nodal
displacements (via equilibrium equations based on elasticity theory), permitting a much
more nuanced understanding of the effects of structural and material irregularities on stress
and strain. In this example, beam theory and FE analysis will yield more or less identical
descriptions of the stress field. With geometrical irregularity and spatial variation in material
properties, which is the rule rather than the exception for bones, beam theory will fail to
characterize absolute stress magnitudes accurately.

a complete depiction of the state of stress and strain in a loaded object. Its promise, from biomechanical and paleontological points of view, is that any morphological configuration can be appraised in this manner. Enthusiasm for this technique in human evolutionary studies has been palpable, with some fossils (Sts 5, OH 5) having been modeled separately by different collaborative teams (Strait et al., 2009; Wroe et al., 2010; A. Smith et al., 2015). This provides an opportunity to evaluate the potential of FEA. These models will not be identical, but if they are robust in their boundary conditions and material assumptions, they will be broadly similar in their descriptions of the stress field. If this is not the case, then, presumably, disagreements among investigators are attributable to differences in model specifications. On the other hand, if the models are convergent in their results, but the adaptive explanations for those results differ, then the method itself is not to blame.

The promise of FEA is that of unprecedented precision and accuracy in describing the stress environment in bone. FEA's precision, however, is often so fine grained that most of the data is effectively ignored in the discussion of results. The sheer volume of data can be intimidating, and, for hypotheses that have been articulated to date, there are usually particular regions of interest that attract researchers' attention. For example, with *in vivo* investigations of the zygomatic arch, sampling three regions can provide a relatively simple and coarse evaluation of a strain gradient. A finite element model will be able to sample strains at a density at least an order of magnitude greater than what is possible using strain gages. This makes the FEA approach preferable—if it is accurate.

While the precision of FEA is uncontested, accuracy is another matter. The issue is that there are a variety of decisions to be made with respect to boundary conditions. What parts of anatomy are restrained? How much force is applied to the model? Where is it applied? What direction is it coming from? Thus there is no guarantee that the model's output mimics what is occurring in a real organism. For this reason, models need to be validated against actual physical (i.e., not purely mathematical) data. This is impossible to do with fossils, owing to the loss of bone material properties in the course of fossilization.

In paleobiological applications, it is agreed that validation tests on organisms for which FEA models and experimental data are available can serve as surrogate tests. There is no formal consensus as to what constitutes a valid model, however, which has resulted in them being judged valid by declaration (Strait et al., 2009; Wroe et al., 2010) or by a criterion of model strains lying within two standard deviations of experimental strains at sampled locations (Strait et al., 2008).[10]

Hominin cranial FEA models all ignore the temporal fascia, which attaches along the superior aspect of the zygomatic arch. This omission presumably explains why the strain gradients in the zygomatics of the models are incompatible with strains measured *in vivo*.[11] Other modeling decisions and their effects have been vigorously contested in the literature, including the inclusion and exclusion of sutures (cf. Kupczik et al., 2007; Wang et al., 2010) and incorporation of the periodontal ligament (cf. Gröning et al., 2012; McCormack et al., 2017). The current state of the art for finite element analyses of fossils is that their local validity should be viewed with considerable skepticism, even if the global form of the models is approximately accurate.

The problems of validation may stem from sources beyond decisions about the inclusion of soft tissue interfaces (noted above). Boundary conditions involving load application are obviously unknowable with respect to fossils, but sensitivity analyses (C. F. Ross et al., 2005) can assess the magnitude of error with respect to such decisions. A sensitivity analysis basically iterates over several alternative sets of boundary conditions to examine how the model output (i.e., stresses and strains) are affected. Bone material properties may represent another source of error, but using modern analogs that could be iterated to document model sensitivity represents a useful, if imperfect, workaround. The detection of secondary bone in microCT surveys of fossils could certainly assist in these endeavors, as they could be represented by pockets of relatively compliant bone within models.

Given the problems noted here, finite element analysis has been oversold as an elixir for paleoanthropological inference. As Rayfield (2007) has argued, FEA is ideally suited for addressing targeted hypotheses of bone morphology. Its principal failing in paleoanthropology has been that whole cranial models have been inspected, in terms of the general distribution of stress under a handful of loadcases, and that characterization has been used to infer broad adaptive significance. This kind of conceptual leap prompted Rayfield (2007:560) to ask, "Should skull shape and mechanical loading be viewed so intimately linked in the first place?" Strait et al. (2009, 2010) essentially resurrected Wolff's law by assigning evolutionary significance to the presence of strain gradients in the maxilla; i.e., the signal is adaptive, rather than architectural. They drew a link between strains and feeding ecology, and what is particularly curious is their conclusion that Sts 5 (representing *Australopithecus africanus*) was durophagous. The difference between ingesting hard foods, rather than any other ones, was never actually explored in their model, but the adaptive inference was made with reference to single biting events;

i.e., there was no attention at all given to a daily or effective stress stimulus. This, in effect, was an application of the paradigm method.

Computational models of evolution are the latest hammer in paleoanthropology's methodological arsenal. These are theoretical simulations that, given a set of assumed parameters, mimic the evolutionary process. The object is not so much the prediction of morphology—although such models have been widely developed in bone biology literature (Pivonka et al., 2018)—but, rather, the estimation of characteristics of hominin populations. Whereas finite element analysis is operating on the tissue or elemental anatomical level, evolutionary simulations function at a population level or higher. Both can employ recursive algorithms to generate data; i.e., these models deploy predefined empirical or logical processes to generate patterns. The output patterns can then be compared, usually incompletely, with the ones based on empirical data. The value of these comparisons is that through such simulations, we can begin to narrow the universe of plausible explanations for observations in the paleontological record. The processes responsible for observed patterns are manipulated *in silico*, in order to understand their influence.

Variability selection (Potts, 1996, 1998) is a provocative model of hominin evolution in which allelic variation for stenotopic versus eurytopic phenotypes explains our successful evolution during times of rapidly fluctuating environmental change. This is reasonable, as long as one accepts the existence of such alleles. In a series of papers, Grove (2011a, 2011b, 2014) simulated the evolution of hominin populations in terms of alleles that had different fitness values across changing environments, including variants that had good-but-not-great fitness across environments but below-average fitness in particular ones. He showed that under constant but incremental environmental change, these eurytopic alleles could indeed spread through populations. This imbued Potts's (1996, 1998) model with an added degree of credibility, through a rigorous testing of evolutionary logic. Because such evolutionary simulations can be iterated across many different sets of initial parameters, Grove (2011b) was also able to further frame the context in which variability selection could operate. Specifically, the persistence of these versatilist alleles was an even-money proposition under an effective population size of as little as 264 individuals. What is important about this simulation approach was how straightforward the evaluation of Potts's (1996, 1998) model appeared to be in terms of its general validity. Grove (2011b) engaged in what is essentially an ecological simulation without needing to delineate specific ecological parameters.

In evolutionary simulations, however, an almost infinite number of permutations of parameters can be iterated, and the stability of the models is ordinarily assessed for a finite number of these. This necessary choice amounts to an unstated rationale: the examined iterations are the best ones available for moving the model into the realism domain (Figure 3.2). Hence the foundational weakness of an evolutionary simulation model is no different from any other; making it operational requires choosing among simplifying assumptions. Grove (2011b) modeled environmental change in two ways: (1) as incrementally fluctuating (via a sine function), and (2) empirically, using paleotemperatures as an environmental surrogate. Both are more or less continuous functions; thus the models treated environments as continuously variable. The surrogate data for "environment" permits this, but from an organism's or population's point of view, it is worth asking whether environments are "chunky"; i.e., objectively discrete in some way. Environmental variance along a transect that includes gallery forest and a savannah (two common vignettes of hominin habitats) suggests that this may be a legitimate question. In a virtual competition of mesic, xeric, and versatilist alleles in a discrete environmental scheme—where changes are sudden, temporally unpredictable, and large—variability selection is somewhat less compelling as a potent process.[12] In this situation, drift gets an outsized influence on allele fixation, but even with large (on the order of thousands) effective population sizes, versatilist alleles do not outcompete their specialized competitors. The point here is that in both models, the environment is imprecise and unreal, and the degree of unreality is a question that needs to be explored.

Simulation models are equally suited to explore bottom-up processes in evolution. Nowlan and Prendergast (2005) produced a simple mechanobiological model, where a set of alleles determined long bone modeling parameters in a population. This model "grew" a population of bones for each generation, with there being a combination of mechanically mediated modeling and a biological baseline of bone growth early on, changing over to completely mechanically determinant growth through adolescence and adulthood. The alleles associated with efficient bones were selected in each generation, with efficiency defined as the balancing of bone mass investment while tracking a target stress interval. Growth in body size was used as a surrogate for mechanical loading, with a modest amount of noise built in. What was instructive about the model was that it generated results confirming the suspicions that bone biologists have held for decades—i.e., bone development did not end up at some obvious mechanically adaptive optimum. What Nowlan and Prendergast (2005) found was that competent phenotypes would evolve, but these did not converge on a theoretical optimum. Their simple model linked morphological change from a developmental scale to a longer-term evolutionary one, using a simple mechanobiological criterion.

Obviously, such models do not tell us what happened during human evolution, but their value is in helping to frame what is possible, or, more importantly, unlikely. These models do not really qualify as hypothesis testers unless they achieve equifinality; i.e., the model's parameters and initial conditions never yield substantively different results. With reference to Sept's (2007) paleontological model scheme (Figure 3.2), these models can run the gamut from general to specific, but to get to the desired space of realism, they need to have data to inform the parameters. Some of this information will come from the fossil record, but at least some of the time, extant analogs will almost certainly have to stand in for paleospecies.

The Relationship of Method to Theory

Functional inference in paleoanthropology is largely an exercise in pattern recognition, carried out primarily through comparative methods. One of the important distinctions to appreciate among methods is that, while they can serve as effective hypothesis generators, in other contexts they are better positioned to be hypothesis testers. Evolutionary simulations have the ability to test certain (i.e., sufficiently detailed) hypotheses in that they can, if iterated appropriately, fail to find an evolutionarily viable outcome from a realistic set of initial conditions. Neontological experiments are important hypothesis testers for relating morphology to performance. When the experiments indicate an unexpected aspect of performance, the opportunity is there for novel hypotheses to be developed.

One area of theory that remains relatively untouched by method or data is the relationship of fitness to observed patterns of morphological variation (Figure 3.1). Functional morphology, as a relevant subdiscipline of evolutionary biology, is premised on fitness having a tangible connection to the details of morphology. Is there a predictable relationship between morphological variation and variance in fitness? This is certainly discoverable (e.g., Miles, 2004; Eklöv and Svanbäck, 2006; Whiteside et al., 2016), but it is challenging to envision how investigations incorporating variables of performance, morphology, and survivorship/fecundity can be fully operationalized in natural field settings, to say nothing of the paleontological context in which inferences of performance and the scope of biological roles rely largely on morphology. Variation in skeletal traits has a nontrivial effect on fitness and is obviously mandatory for understanding adaptive trends in hominin evolution, but in this case, necessity has not yet given birth to invention.

Since biological roles cannot be discerned independently of morphology in the paleontological context, an incremental approach is called for. A successful one will involve collaboration among morphologists, field biologists, and geneticists. S. Arnold's (1983) call for laboratory investigation of performance, coupled with field observations of the context of performance, is widely appreciated outside of paleoanthropology (Kingsolver and Huey, 2003). Understanding the genomic basis for morphological variation is vital for understanding the evolution of modularity (Cheverud, 1984), as well as the evolvability of morphology (Klingenberg, 2005). Wagner and Misof (1993) made the distinction between generative and morphostatic constraint: they are both developmental, but they differ, in part, in terms of timing. The generative constraints are operative earlier, involving morphogenetic interactions, while the morphostatic limit variation is due to stabilization of patterns later in development. The distinction is valuable, because generative constraints are hardly absolute across taxa, and morphostatic ones have an overarching influence on conservation of adult morphology. This runs counter to the general idea that the buffering of morphology necessarily diminishes as a function of time during development. This is yet another challenge to functionalist logic.

How do these qualifications actually help the researcher staring at the fossils on her lab bench? A pragmatic overall strategy for functional inference in paleobiology is managing the tension between the necessity of informed speculation and the reining in of unbridled imagination. What qualifies as unbridled may be resolved by some protracted disciplinary argument, but a more immediately effective tactic is to use comparative data, analogs, or models as foils. There are several historical examples of the value in this approach, and I will briefly review two here. In each of these cases, had developmental data been recruited to understand the morphogenesis of the traits studied, the initial conclusions would probably have been dismissed, or at least qualified.

Using fossils to infer the details of neuroanatomy has been practiced since Dart (1925) contemplated the meaning of the Taung endocast. Kay et al. (1998) suggested that resolution of the apparently intractable debate over the appearance of language in human evolution could be achieved by assessing differences in hypoglossal nerve size in modern and fossil taxa. The logic of their approach was perfectly clear: humans must have fine motor control of the tongue to facilitate the phonetic precision of spoken language. The hypoglossal nerve, being the principal motor supply, must consequently furnish more motor units, which means more axons, to the tongue. Given this reasoning, humans ought to have relatively large hypoglossal nerves, compared with other animals. The hypoglossal nerve, being a cranial nerve, arises directly from the brain and, to get to its target organ, must exit the skull, which it invariably does by traversing the occipital through the canal that bears its name. Kay et al. (1998) reasoned that a large hypoglossal canal in a skull betrays what once was a large 12th cranial nerve running through it.

In terms of both absolute and relative size, Kay et al. (1998) found that human hypoglossal canals are large when evaluated by a narrow comparative sample of chimpanzees, bonobos, gorillas, and a few fossil hominins. This sample constituted what is typically regarded as the appropriate phylogenetic bracket in hominin paleobiology. Because Kabwe, Skhūl V, and La Ferrassie hypoglossal canals fell within the size-adjusted human range and outside that of australopiths and apes, the obvious conclusion was that late Pleistocene *Homo*, including Neanderthals, had the capacity for speech. A long-standing debate on the origins of language appeared to be on the brink of resolution.

DeGusta et al. (1999) undermined this conclusion by presenting data suggesting that hypoglossal canal size and nerve caliber (or the number of axons within the nerve) are uncorrelated. They also expanded the comparative sample beyond the hominid clade to include numerous other primate species, finding that relative canal size can be large in several species (all nonspeaking) of nonhuman primates. In a counterargument to their rebuttal, Jungers et al. (2003) noted that the size correction employed by DeGusta et al. (1999) was flawed, thus explaining relatively giant hypoglossal canals in lemurs.[13] But even this reanalysis, as well as a comparison of chimpanzee and human hypoglossal nerve sizes, indicated that there was not empirical support for exceptional hypoglossal nerve or canal size in humans. The absence of correlation between nerve and canal size was also seemingly explained by the fact that the bulk of the canal's volume is not occupied by the nerve, but by the venous plexus that is typically found within it.

This perfectly reasonable hypothesis depends on the validity of what appears to be an equally reasonable premise. Thus examination of these premises is obviously crucial. The underlying philosophical position of the original argument is essentially functionalist: the space through which a nerve passes is determined by the nerve itself, providing a harmonious relationship between the passageway and the thing running through it. Yet there is a cause-and-effect undercurrent here that is unnecessary. While the nerve must traverse the canal, there is no indication that it determines canal development; one can only be sure that the nerve needs to exit the skull unimpeded, or else natural selection will spring into action. The structuralist view, retrospectively, is that the bone cells lining the canal are following morphogenetic rules that are not inextricably responsive to the presence of the nerve.

For a second example, consider another long-standing question in paleoanthropology: whether the basal hominin descended from a knuckle-walking ancestor. The phylogenetic proximity of gorillas and chimpanzees to humans more or less compelled consideration of this hypothesis (see Chapter 4). Richmond and Strait (2000) declared the question solved, based largely on a multivariate analysis of carpal features among living and extinct hominids. Their claim was not made without a biomechanical explanation. The authors noted that in the extended chimpanzee wrist, which is thought to typify knuckle-walking behavior, the radiocarpal and midcarpal joints achieved a "close-packed" position, owing to peculiarities in chimpanzee radial and carpal morphology. The demonstration of knuckle-walking affinities was also articulated as the relationship of species morphology in multivariate space. Comparisons were informed by a phylogenetic bracket of Great Apes, modern humans, australopiths, and idiosyncratic representatives of the Cercopithecidae and Ceboidea (*Erythrocebus, Papio,* and *Alouatta*). Subsequent missives in the literature (Corruccini and McHenry, 2001; Dainton, 2001; Richmond and Strait, 2001; Richmond et al., 2001) entailed arguments about the proximity of taxa to one another on canonical axes. Beyond this, the dispute revolved around discrete morphological features of the distal radius shared by African apes, which appeared in australopiths as well. Richmond et al. (2001) argued in favor of knuckle-walking ancestry for humans based on (1) a comprehensive accounting of forelimb anatomy, (2) the unlikely possibility that knuckle-walking would have evolved independently in African apes, and (3) the inadequacy of alternative models of locomotor adaptations in hominin precursors. On balance, they concluded, their analysis "allows us to reject scenarios that rely on a strictly arboreal ancestor" (Richmond et al., 2001:98).

The ingredients for the standard functional inference were all present: (1) a unique association between an anatomical feature and a behavioral capacity, (2) a demonstration of morphological proximity of fossils to the taxa possessing the features in question, and (3) an invocation of phylogeny to countermand alternative interpretations. The discovery of *Ardipithecus* largely dispensed with the above conclusion (see Chapter 4), but its weaknesses were clear enough, given a modest degree of skepticism (Dainton, 2001).

A unique association of anatomy and behavior is not the same thing as a demonstration of adequate or competent performance, in an absolute or relative sense, but the inference of performance is usually not difficult to draw theoretically, even if empirical support is lacking. The greater issue, pragmatically speaking, is that of unique association, including whether the search for exceptions has been adequate. Kivell and Schmitt (2009) simply expanded the primate sample and found putative knuckle-walking traits in taxa that do no such thing. They also discovered that certain alleged knuckle-walking features were absent in gorillas, thus nullifying the diagnostic status of the trait in question. This "impossible" state of affairs is explained by the fact that knuckle-walking is distinct among gorillas and chimpanzees. The hyperextended carpal postures in chimpanzees are not utilized by gorillas, who do most of their knuckle-walking on the ground, rather than in trees. Kivell and Schmitt (2009) also recruited ontogenetic data (cross-sectional sampling of age series) to undermine the alleged diagnostic status of some of the features linked to knuckle-walking. Given behavioral data, such features should appear earlier in gorilla ontogeny than in *Pan*, but these traits often emerged later and less saliently in gorillas, if at all.

In retrospect, there are several reasons why the declaration that australopiths descended from knuckle-walkers is problematic. Once again, this has more to do with implicit assumptions about the information content of data than measurement of the variables themselves. First, the argument over multivariate proximity may not have much to do with locomotion per se. Morphometric proximity is a systematic partitioning of variance, and the juxtaposition of taxa in that multivariate space has no necessary connection to function or efficiency.

The second prong is that the independent acquisition of knuckle-walking in African apes is an affront to parsimony. This rationale is hard to explain, other than as an uncritical invocation of the principle. The concept of parallelism is based in the idea that, given a relatively recent common ancestry, descendant taxa have a shared set of anatomical complexes, underlain by morphogenetic processes, that result in finite variations in skeletal form. Homoplasy is easy to find, once one decides to look for it (Wake, 1991; Lockwood and Fleagle, 1999; B. Hall, 2007); it is also integral to a structuralist paradigm. Moreover, Kivell and Schmitt (2009) argued, on anatomical-behavioral grounds, that there are actually distinct locomotor adaptations operating in *Pan* and *Gorilla*, while our locomotor classificatory scheme implies that knuckle-walking is monolithic.

Finally, and certainly not confined to the cases at hand, paleoanthropological "insights" can derive legitimacy from exposing holes in alternative models. The irony is that general models, which situate their arguments above the scale at which they can be meaningfully interrogated, immunize themselves from tests. Models that specify the details of performance, as well as how those capacities come into being developmentally, are better equipped to understand the evolution of the associated morphology. Similarly, a focus on the particulars of behaviors dependent on morphology permits refutation by experiment or comparative examination. This is what models in paleoanthropology should aspire to.

The preceding examples underscore that neontological data provide the necessary framing for drawing inferences in paleobiology. In that vein, there is also a need to bring process-focused methodology and data more to the fore in functional morphological studies. Both developmental and evolutionary models can direct and inform research that may succeed in identifying constraint in concrete morphological terms. The reflexive caveat at the conclusion of papers in which the prescribed form-function coalescence failed to materialize is that some sort of phylogenetic or developmental constraint had prevented it. The subtext is that the fit of theory to observation should be better, but if natural selection is ubiquitous, and constraint is insurmountable most of the time, then what is observed is just the best evolution could do, given the vagaries of the focal trait's history. If we understand and expect that the adaptive capacity and plasticity of the skeleton is finite, we ought to cease being surprised that precise matches of form to function do not occur. In cases of excellent fit, we should always explore the extent to which function is adequately understood, the reason being that it is all too easy to retrofit function to form in the context of fossils.

Direct observation of developmental processes in fossils is not possible, so evolutionary processes have to be discovered indirectly, through the analysis of pattern. Bone does leave traces of process, however, in the microstructural signatures of primary and secondary bone. These are preserved in fossils and can be used to infer bone metabolic activity (Bromage, 1989; Daegling and Grine, 2006). A critical question for systematic surveys of fossil material is whether microCT or syncrotron imaging permits visualization of microanatomy in fossils. Paleoanthropological applications to date using microCT look promising (e.g., Mednikova et al., 2013; Bradfield, 2018), and Sanchez et al. (2012) demonstrated feasibility using syncrotron tomography.

Functional inference is problematic in paleoanthropology, because uncertainty is intrinsic to the enterprise. If the problem is not simply a dearth of material, it is worthwhile to ask whether methods are inadequate or theory is misguided. The latter query might be beyond our discrimination—of course we are satisfied that we have a handle on the evolutionary process, especially since we have developed an appreciation of what we do and do not know (Ungar, 2006). We are at the point now where a purely mechanically utilitarian view of bone is untenable.

To judge whether methods fall short, it is necessary to specify what they are employed to do. If methods in functional morphology are intended to document variation and its relationship to performance, then our approaches can be equal to the task. If the methods are supposed to provide insight into selective regimes and fitness, then their efficacy is much murkier. This latter problem, moreover, is not unique to paleoanthropology.

The growing realization that adaptive or behavioral inference is not straight-forward indicates that the process of situating functional morphology within the broader discipline of evolutionary biology is still being worked out, and that where the field is going will be firmly centered in a structuralist philosophy. Our more nuanced appreciation of adaptation in the past half century indicates that this is taking place. Methodologically, our approaches are adequate for meeting the foundational goal of understanding how morphological systems work. The current issue is to discover how the evolution of anatomical complexes facilitates or re-strains the universe of possible behaviors. If the overarching question inspiring research design is, "Why is morphospace discontinuous?" then we will incrementally but certainly discover the nature and scope of constraint.

Measuring performance does not inevitably lead to the discovery of behavior itself. This cannot be credibly done without data that are independent of the morphology-performance-fitness inference. In practice, functional morphological data are indiscriminately recruited to inform ecological and evolutionary domains. As the examples reviewed in this chapter indicate, there is an unjustified leap from morphological distance (a methodological endeavor) to evolutionary inference (a theoretical judgment). The quantification of morphology does not lead seamlessly toward understanding its evolution.

Chapter Four

Bipedality

D arwin raised the question of what advantages accrue to the adoption of bipedal posture. His speculation that the freeing of hands—enabling the manufacture and use of tools—precipitated bipedality seemed to be so sensible that the question apparently was solved. As the advent of bipedality has been pushed back to well before the earliest good evidence for tool use, that obvious explanation is more difficult to sustain. The focus in paleoanthropological literature prior to 1978 was largely concerned with the adaptive explanation for the adoption of bipedality, rather than how, as a matter of anatomical transformation, the transition from some other form of locomotion to bipedality occurred. That changed after the discovery and description of *Australopithecus afarensis*, when that treasure trove of postcranial fossils sparked a decades-long, entertaining, and acrimonious series of arguments over the nature of bipedal progression in that taxon. This debate was illuminating with respect to functional morphological theory and practice in paleoanthropology. On one side was what I will refer to as the *efficiency camp*,[1] spearheaded by Owen Lovejoy, which used biomechanical analyses to argue that *Au. afarensis* practiced a form of bipedality as functionally similar to and perhaps as economical as our own. The *compromise camp*,[2] led by Jack Stern and Randall Susman, employed biomechanical analyses to argue that the kind of bipedality in *Au. afarensis* was fundamentally distinct from that of modern humans, being relatively inefficient energetically. The implication of these incompatible conclusions at the behavioral level centers on the question of arboreality. Had it been forsaken entirely, or was it an adaptive component of early hominin ecology?

The Ecological Question

Explanations of the origins of bipedality have, unsurprisingly, made use of top-down scenarios that considered hominin interactions with local landscapes, biotas, and conspecifics, all necessarily described in generalities (Hardy, 1960; Morris, 1967; Ardrey, 1976; Lovejoy, 1981). All of these ecological arguments leave an alarming latitude of interpretive license (Langdon, 1997), and most scenarios can be made believable by the judicious application of convincing threads of logic containing uncontroversial general assumptions. Explanations regarding what qualifies as the critical historical adaptation are complicated by the possibility (infrequently explored) that many scenarios are internally consistent, not just on their own, but also in combination with other scenarios. If early hominin bipedalism incurred advantages in open-country foraging—for example, with respect to seed-eating (Jolly, 1970) or provisioning (Lovejoy, 1978)—thermoregulatory benefits would accrue incidentally (Wheeler, 1991; see also Ruxton and Wilkinson, 2011).

Top-down scenarios outline an ecology of bipedalism, and the biomechanical particulars of bipedalism may or may not have any role in such explanations. What matters is that there must be an external environmental driver that initiates a novel selective regime. An example would be occupation of open-country habitats by a hominin species. Arboreal behaviors and the morphology that permits them can be actively selected against if (1) there are no trees, and (2) terrestrial locomotion is sufficiently compromised by the once arboreally adapted anatomy. The details of morphological change are less important, because the overarching problem, under this scenario, is identifying the environmental context of selection for bipedalism.

Despite this facile example, the ecological context does matter for a general justification of hypotheses of biomechanical and functional performance. The efficiency camp argue that selection for bipedality is, in some sense, absolute. Terrestrial progression on two limbs requires a total commitment, because inefficient bipedality negates its sole advantage prior to tool-making culture—the long-distance transport of food, which is tied to the reproductive strategy of provisioning (Lovejoy 1981, 2009). Trees are present, but they matter less, because being good at climbing and moving around in them is too costly (Latimer, 1991).

Susman et al.'s (1984) approach was to pose the question, If you are a 1.3 m tall, 30 kg hominin, can you afford to forsake the trees, given your higher-up compatriots in the African food chain? The inefficiency of bipedality is understood here in relative terms. If *Au. afarensis* is better in the trees than *Homo* and better on the ground than *Pan*, then what the fossils represent is a creature well adapted to the early hominin niche.

Both justifications make sense, but they also beg the question. While the use of trees by early hominins for foraging or safety is certainly desirable to know, there may be no way to learn if this was the case from fossil material. A more tangible path to resolution is to inquire as to the nature of bipedal efficiency and how this does and does not conflict with competence in the trees. The fact that most primates utilize trees for foraging and protection (*Theropithecus* and some *Gorilla* being exceptions) only resolves the argument if arboreality is deemed compulsory, because of their primate status.

The Energetics Question

One reason there is little argument that bipedality must have evolved under intense selection is that, by the optics of performance, it seems to be such a bad idea. With respect to generalized primate quadrupedalism, it is both unstable and slow. An energetic advantage of bipedalism is realized through the extended hip, extended knee walking gait practiced by modern humans. Under these conditions, energetic costs are relatively low, because muscular effort is used not so much for propulsion, but rather for postural control and management of the exchange between potential (gravitational) and kinetic energy. The instability of bipedalism then becomes an advantage, in that gait is initiated by a controlled fall forward. This energetic advantage over quadrupedalism is mostly lost when running (C. R. Taylor and Rowntree, 1973; Bramble and Lieberman, 2004; Pontzer, 2017a, 2017b).

Lovejoy (1975, 1978, 1981, 1988) and his colleagues (Lovejoy et al., 1973, 2002) focused on energetics as the guiding principle for understanding the evolution of human locomotion. The facultative bipedality practiced by nonhuman primates is characterized by a compliant gait (Schmitt, 2003), in which the knees and hips never achieve full extension, and both vertical oscillations of the center of mass and ground reaction forces are relatively small. This type of bipedality actually enables faster walking speeds, but the energetic costs are high.[3] This observation means that if a compliant gait characterized the initial form of bipedality, selection based on energetic savings is unlikely. Chimpanzees adopt a compliant gait when walking bipedally, and, compared with their quadrupedal walking, there are no energetic savings (Pontzer et al., 2014).

The inefficiency of the compliant gait is taken as *prima facie* evidence that, as a mode of locomotion, it could not have been the target of selection in hominin evolution (Crompton et al., 1998). This statement must be distinguished from one claiming that a compliant gait could not have evolved. Schmitt (1999) noted that compliant gaits are a peculiar characteristic of primate locomotion in general, and there may be neuromuscular constraints that operate generally among the order that predispose its members to utilize them. In any case, the position of the efficiency camp is that even if a compliant gait is behaviorally possible, it is ecologically intolerable over the long term. Consequently, a punctuated mode of locomotor change is required if a compliant phase was involved in the evolution of bipedality.

This need not imply a symmetry of explanation, in which the arguments of the compromise camp depend on a gradualist, monotonic mode of morphological change. The position that early australopith bipedality was unlike that in later *Homo* is not incompatible with any particular assumptions of evolutionary tempo or mode, but it does adhere to a principle of mosaic evolution, as opposed to some form of orthogenesis. Whether this serves to contrast the two schools of thought depends on what "mosaic" is supposed to mean. If we allow for a conventional definition, where different traits or features can evolve at distinct rates, then it is fair to say that both camps subscribe to this idea. On the other hand, if, by mosaic evolution, we are talking specifically about responses to distinct selective forces, there are clear differences in the two camps' presumptions and outlook. One school of thought is panselectionist, and the other sees ontogenetic or phylogenetic constraints as ruling out certain combinations of morphological features.

In terms of whether a thermodynamic criterion is adequate to settle the debate on the nature of early bipedality, the crucial question is what the costs of inefficiency are in terms of fitness. This reductionist tactic ultimately depends on the ecological inference. If foraging behavior is such that long-distance travel is trivial or nonexistent, the fitness advantage of striding bipedalism is reduced. Similarly, at some point arboreal locomotion becomes selectively unimportant as the proportion of terrestriality increases. In early hominins, climbing would have been more expensive energetically than walking (Hanna et al., 2008), so presumably this question can be modeled by energetic criteria. Multiple sets of boundary conditions (e.g., assumptions regarding landscape use and feeding ecology) would need to be considered.

The Precursor Question

If, as evolutionary biologists will readily concede, phylogenetic heritage influences morphological evolution, then the locomotor repertoire that preceded the adoption of bipedality is something we want to know. There are at least two reasons for this. First, from the standpoint of being diligent in conferring adaptive or exaptive status to traits, we need to know which traits were in need of transformation and were thereby subject to selection, and which ones may have already been present. Second, do certain locomotor repertoires predispose shifts to bipedality more readily than others? Perhaps, the reasoning goes, bipedality evolved because it was the easiest or only option after descent from the trees.

The notion of bipedalism's antecedent form being informed by imposition of a phylogenetic bracket was understood long before the concept was formalized. Prior to the emergence of molecular phylogenies, however, there was room to argue about the topology of the bracket. This is no longer the case, and there is little reason to suppose that the form of the bracket is going to change in the future: *Pan* is our sister taxon, to the exclusion of *Gorilla*, *Pongo*, and *Hylobates*, in that order. Our task is to use the fossil record and locomotor anatomy data from extant hominoids to reconstruct which morphology resides at the node joining *Pan* and *Homo* (Figure 3.5).

Parsimony would dictate that this reconstruction would represent a knuckle-walker, because, failing that, this relatively rare locomotor pattern would have had to evolve twice. As noted in the previous chapter, this represents an insane devotion to the principle and reflects the strange phobia of homoplasy in paleoanthropology. If parallelism is a valid concept, then an equally plausible alternative is the independent evolution of knuckle-walking, given two closely related, large-bodied apes with high intermembral indices,[4] long fingers, and a lifestyle requiring significant time spent on the ground and in the trees. Given the presumed proximity of *Ardipithecus* to the node linking *Pan* and *Homo*, Lovejoy et al. (2009) suggested that *Ardipithecus* represented a decent model for the last common ancestor between humans, on the one hand, and chimpanzees and bonobos, on the other.

This reconstruction painted *Ardipithecus* as a chimera, unlike all extant hominoids in that this genus purportedly lacked specializations for knuckle-walking, vertical climbing, suspension, and (to no one's surprise) brachiation. In effect, the precursor to bipedalism is a generalized capacity to climb and bridge in an arboreal setting, with above-branch quadrupedalism the preferred means of moving over longer distances. If this is an accurate representation, then the phylogenetic bracket is more misleading than helpful. One implication of this particular reconstruction is that the locomotor repertoire at the critical node is occupied by a facultative biped. Lovejoy et al. (2009) are explicitly agnostic in this regard, but given facultative, if infrequent, bipedality in living *Pan* and obligate bipedality in hominins, there is scarcely any good reason to doubt some form of bipedality in the last common ancestor.

The break between this account and historical ones contemplating the precursor to bipedality is striking, if for no other reason than the *Ardipithecus* reconstruction apparently renders extant nonhuman hominoid locomotor adaptations irrelevant for understanding bipedalism's origins. This troubled Pilbeam and Lieberman (2017:28) sufficiently to prompt them to doubt the reconstruction, because it would require that "almost all postcranial similarities among the extant hominoids must be homoplastic." A common theme in prior reconstructions was the role that orthogrady may have played in the emergence of bipedality. Orthogrady is a postural rather than a strictly morphological concept, although it readily distinguishes monkeys from apes in several anatomical details. The basic idea is that orthograde habits predispose a population to terrestrial bipedalism. Tuttle (1981) invented a hypothetical hylobatian ancestor that was so committed to orthogrady that it was bipedal in the trees. Stern (1975) offered that a hominoid with hylobatid-like limb proportions may have been forced to adopt bipedality on the ground out of biomechanical necessity. While intuition suggests to many that a long-armed quadruped would bear relatively more weight on the hindlimb, Stern (1975) argued mathematically that the opposite is true. Since a high intermembral index causes more load to be borne by the forelimb, he reasoned that quadrupedalism would be intolerable in a previously suspensory-adapted primate.

Most reconstructions of the precursor to hominin bipedalism necessarily rely on comparison of osteological features and general body proportions, such as crural or intermembral indices. But electromyographic evidence has been used to argue that climbing is linked to bipedalism by virtue of muscle function. Fleagle et al. (1981) observed that vertical climbing involves phasic activity in several hindlimb muscles, reminiscent of that seen in bipedality. Climbing, therefore, provided a biomechanical substrate for the acquisition of bipedality. Under this scenario, the most appropriate model for pre-hominin locomotion is the orangutan. This model has appeal for two reasons. First, it does away with the need for a brachiating phase preceding hominin emergence. Second, it provides a behavioral justification for linking bipedality and arboreality early on in hominin evolution, without needing to have the creatures walk in the trees.

Climbing, of course, is compatible with the classical locomotor types in primates; i.e., brachiation, suspension, or quadrupedality. The two genera of knuckle-walkers are quite competent climbers, as well. In fact, one component of the argument in favor of knuckle-walking ancestry for the earliest hominins is that climbing is in no way precluded.

The phylogenetic bracket encourages the idea that hominoids are a clade united by certain features (shoulder mobility, orthogrady, long cheiridia) that define a functional gestalt. Even the curiosity of knuckle-walking can be understood as a consequence of a commitment to climbing and suspensory behaviors: flexing relatively long digits make for easier progression on the ground (Tuttle, 1969). The fossils, however, offer meager support for the idea that hominoid specializations were present at the origin of the clade. *Sivapithecus*, having an unambiguously pongid face, shows few of the skeletal correlates of suspension in *Pongo* (Madar et al., 2002), and gibbons are so autapomorphic postcranially as to be practically unrecognizable in the fossil record in the putative time range of their origin (Jablonski and Chaplin, 2009).

Lovejoy et al.'s (2009) reconstruction of locomotion in *Ardipithecus* was remarkable because the morphological pattern observed was unpredictable, but such discoveries provide opportunities to understand epistemological lacunae. One illusion that persists in paleoanthropology is that we have a decent understanding of the pattern of hominin evolution. Another is our inability—or unwillingness—to recognize contexts when uncertainty prevails (the "black swans" of Taleb, 2007). The morphology of WT-17000 was a completely unanticipated variant of *Paranthropus*; Wood and Collard (1999) explained how the discovery of KNM-ER 1470 induced cognitive dissonance among the paleoanthropological community. This recurrent theme of complete surprise reflects an unconscious commitment to both essentialism and parsimony. Typological thinking, not vagaries of evolution, explains the astonishment that greets the discovery of each new hominin species.

The history of thought on the locomotor profile of protohominins is anchored on some variation on hominoid themes of orthogrady, suspension, climbing, or what could be collectively called "antipronogrady" (Stern, 1975). This rationale is steeped in principles of parsimony and phylogenetic constraint. *Ardipithecus*, however, provides an example in which these principles fail to retrodict locomotor behavior (White et al., 2009; cf. Pilbeam and Lieberman, 2017). Understanding the role of constraint in hominid evolution is badly needed. If constraint is to be of any explanatory value in evolutionary morphology, it must operate so that selection can only produce a subset of competent morphological complexes in a particular lineage or clade. This is the morphogenetic perspective eventually articulated by the efficiency camp (Lovejoy et al., 1999). It has the flavor of "genes are destiny," except this concept of the gene is considerably more nuanced than the beanbag concept, where genes have anatomical specificity. Common developmental pathways produce homoplasies, and common ancestry maintains synapomorphies, but the features are morphometrically similar via either process.

Under morphogenetic constraints, selection operates but is not solely governed by choosing among a continuum of performance values, because the morphology underlying performance may be discontinuous. The appeal of the historically popular, antipronograde bauplan of the Hominoidea is that, via selection, the path of least resistance is bipedality. The outcome is preordained, because selection along alternative lines will not persist, owing to biomechanical incompatibilities. Under a perspective in which not all developmental paths are possible, biomechanical solutions are finite and, most likely, suboptimal.

Same Fossils, Different Functions: Compromise versus Efficiency

The historical debate over locomotion in *Au. afarensis* is, to some degree, a contemporary struggle between functionalist and structuralist paradigms. Alberch's (1980) thesis that morphological evolution was biased in certain directions was taken up by Lovejoy et al. (1999, 2009) as a late salvo against the functionalist sensibilities of the compromise camp. Yet this is an oversimplification in explaining the differing conclusions between the research programs. As developed through specific examples below, the primary distinctions in the two approaches are (1) what to do with primitive characters, and (2) how these should be weighted in an analysis. The cladistic framework is more important here than is generally appreciated. The compromise camp understood but largely ignored it, while the efficiency camp used this information as a shorthand invocation of developmental or genomic constraint. Despite recognition of the reality and potential influence of constraint in morphological evolution, neither camp operationalized the concept in any significant way vis-à-vis actual characters. It was used instead as a rhetorical mace.

Ward (2002:188) suggests that beyond this difference in emphasis, interpretive differences arose because "these two sets of researchers are asking different questions." The bilious exchanges in the literature suggest otherwise. Ward believes that the efficiency camp was primarily concerned with identifying the products of selection (i.e., historical adaptations), while those in the compromise camp

(Stern, Susman, and their colleagues) were interested in behavioral reconstructions independent of their status as aptations of various types. In fact, the two sides were preoccupied with both. The anchors that explain the differences epistemologically are the ecological assumptions about what makes a competent hominin. These make character states and morphological distances mean different things.

Methodological errors have been blamed for interpretive differences. Undoubtedly these exist to some degree, but probably not to the extent where the mistakes incurred account for the stark differences in locomotor reconstruction. As an example, measurements of condylar length ratio on the original of AL 333-4 (Lovejoy et al., 1982) and a cast of that specimen (Susman et al., 1984) differ by less than 1%; the ensuing argument about the significance of the ratio had to do with the validity of using a different definition of this ratio in terms of landmarks. In other cases, more substantial errors are likely. The most obvious example is the AL 288-1 pelvis, which suffered postmortem damage. This fact provided considerable cover for creative license in the revealed reconstructions.

Errors of omission are more of a problem. The cross-sectional area of the calcaneal tuber in *Au. afarensis* is outside the range of African apes (Latimer and Lovejoy, 1989), except that if AL 333-37 is included, then this specimen falls on the mean of *Gorilla* and outside the human range (Susman and Stern, 1991). The difficulty rests on an explanation of functional significance, based on the morphometric variable. Absolute size is measured, and it is supposed to reflect load-carrying capacity in locomotion. This is valid, provided the calcaneal area can be scaled to body mass (or a worthy surrogate), but no such attempt is made. The unstated assumption here is that area is proportional to loads experienced in the specimens.

This underscores a problem that occurs with respect to the linkage of theoretical axioms to the informational content of the variables themselves. Latimer and Lovejoy (1990b) examined proximal pedal phalangeal "articular sets" in *Au. afarensis* and compared them with those of African apes and humans, in order to establish "kinematic and kinetic equivalence" in the hominins, to the exclusion of apes. Support for this position was provided graphically, using the criterion of the planar tangent to the joint surface, defined in a sagittal plane (Figure 4.1). For kinematic reconstructions, these visuals may provide information on maximum possible excursions, but there is considerable slop in moving from osteological range of motion to *in vivo* capacity. Manafzadeh and Padian (2018), through exploration of joint mobility with intact supporting ligaments in an avian, made a

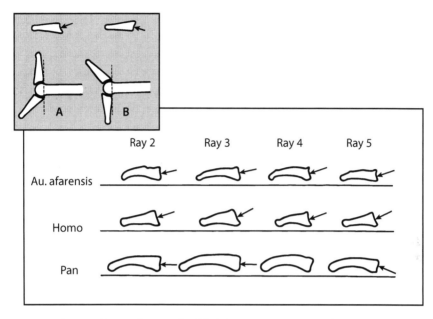

Figure 4.1. The articular set of proximal pedal phalanges, as reconstructed by Latimer and Lovejoy (1990b). The *arrows* define articular orientation as the tangent normal to the articular surfaces, indicating a qualitative similarity between *Australopithecus afarensis* and modern humans. The joint configuration is used to reconstruct the range of motion at the metatarsopahalangeal joint (*inset*), with hominids (A) exhibiting enhanced dorsiflexion and apes (B) having greater plantarflexion capability. Duncan et al. (1994) questioned the reliability of Latimer and Lovejoy's (1990b) method for reconstructing the articular set, and the latter provided no validation of their method for determining the set or its kinematic consequences.

strong case that joint mobility is not reliably inferred from consideration of osseous articular morphology alone. When this is explicitly investigated, actual range of motion is found to be quite restricted, relative to what appears to be osteologically possible, given articular geometry (Cobley et al., 2013; P. Arnold et al., 2014). Neontological data on range of motion may nonetheless productively inform paleontological inferences (Hutson and Hutson, 2013). Reconstruction of kinetic variables is bound to be equally challenging if joint force magnitude and direction have to be estimated exclusively from osseous morphology.

Another example from Latimer and Lovejoy's (1990b) study sought to establish whether *Australopithecus* truly had "long" phalanges. To express this in absolute terms is fairly meaningless in a comparative context, so the authors provided four indices of phalanx length to argue that—on balance—*Au. afarensis* had considerably shorter toes than those of African apes, in some cases falling within the range of *Homo*. Phalangeal length was evaluated relative to (1) femoral length, (2) femoral head diameter, (3) S1 centrum width, and (4) humerus length. The morphometric meaning of these contrasts was that *Au. afarensis* phalanges were short, given the sizes of nonpedal elements. What this indicates about pedal function or adaptation is by no means clear. To understand the mechanical import of phalangeal length requires consideration of its function within the foot skeleton as a whole. With respect to performance, the ratio of phalangeal length to humeral length provides no clear insight. Nevertheless, based on these indices, Latimer and Lovejoy (1990b:22) concluded that "stabilizing selection for long digits had clearly been superseded by directional selection for shorter toes."

Susman et al. (1984) also published comparisons of proximal pedal phalangeal length relative to femoral head size, but they included Lucy (AL 288-1), in addition to AL 333-115. Bonobos represented African apes in this comparison, but the relative position of the australopiths was no different from that found by Latimer and Lovejoy (1990b). The meaning of the intermediate position of the fossils was considered in the context of climbing ability. There was no overt appeal to historical adaptation here; indeed, the focus was clearly on biological roles.

There is considerable faith that metatarsal heads provide a locomotor signal, but what is signified depends on who you ask. It is worrisome that the same skeletal elements can yield disparate functional inferences, but this is enabled when the correspondence of anatomy with actual function is not directly tested. In the case of the metatarsophalangeal (MTP) joints, Latimer and Lovejoy (1990b) drew their "almost-human" conclusion for australopith joint configuration from examining anatomy (specifically, two-dimensional geometry) in a sagittal plane. Susman et al. (1984) offered an equally plausible "ape-like" condition from looking at metatarsal heads in a coronal projection. It solves nothing to simply claim bias on the part of one or both camps. What counts are the biomechanical consequences of the anatomical variation (Lovejoy, 1975), and no neontological performance data were brought to bear on the question by either camp.

Those kinds of data could resolve to what degree articular geometry determines mobility and load accommodation *in vivo*. One intriguing finding in *Papio* is that angular excursion of the MTP joint is similar with respect to both quadrupedal and bipedal progression, with indications of slightly greater excursions bipedally (Berillon et al., 2010). This underscores two ideas that should temper functional inferences from fossils: (1) mobility is not strictly confined to the ranges experienced during typical activity, and (2) similar excursions can be competent for very different (categorically, but not kinematically) locomotor activities.

This is the stuff of headaches for the functional morphologist: identifying that an anatomical complex is biomechanically suited for a particular function does not rule out that the possibility of other functions. Typically, the response to this is phrased in terms of fitness and a thermodynamic adaptive criterion. It does not allow one to dismiss other behaviors as possible, but only to label them as ineffi-cient. In this sense, the efficiency camp enjoys a strategic advantage, in that their exclusive focus on adaptation allows alternative functions to be classified as some-thing else. This is the teleonomic perspective advocated by Williams (1966): other functions are incidental outcomes of selection that produced the adaptation. This implies that incidental functions have no evolutionary significance, and, in any case, ignoring alternative functions fails to provide a complete accounting of be-havior. This is unproblematic only if identification of historical adaptation is the singular objective of functional analysis.

Information content of morphological variables relative to performance is ob-viously critical, but variable definition and measurement will condition such con-tent at the outset. This leads to an emphasis on measurement technique as a point of argument, but, as illustrated in the following examples, the discussion rarely alights on the important question of whether what is measured reflects the metrics of performance. This engenders a curious and unfortunate situation in which the particulars of mensuration are laid out and a broad ecological implica-tion is claimed, but the proximate effects of performance on behavior—the criti-cal objects of selection—are routinely bypassed.

The Pedal Rays

There is one point of agreement concerning the osseous anatomy of the hallux of *Australopithecus afarensis*, which is that the first ray could not be abducted in *Au. afarensis* to the degree observed in chimpanzees. Areas of contention center on (1) conformation of the tarsometatarsal joint, (2) articular set of the first MTP joint, and (3) what the Laetoli footprints might reveal about function of the first ray. Latimer and Lovejoy (1990a) described the orientation of the medial cunei-form's distal articular surface as lacking an oblique medially facing component, which was reasonably interpreted as indicating a reduced degree of hallucial ab-duction. Stern (2000) countered that this surface was curved, relative to the con-dition in *Homo*, which—if based on Latimer and Lovejoy's (1990a) figure—is in-disputable. This suggests some capacity for abduction or adduction, on the principle that articular geometry must balance the incompatible objectives of mobility and stability. That is, if the human joint geometry is stable, then the australopith condition is not and is prone to relatively large local stresses within the joint. Oddly, human first cuneiforms, as figured by Latimer and Lovejoy (1990a), had a greater medial orientation of the distal joint surface than those of *Australopithecus*. In any case, the means by which this medial set was determined is unclear, and it quite obviously depends on the proper orientation of the cuneiform with the navicular. Berillon (2004) suggested that the osseous anatomy indicates limited abduction of the hallux in *Au. afarensis* by an African ape standard, but a less stable configuration than *Homo* under bipedal locomotion.

Duncan et al. (1994) remarked that the demonstration of articular set of the metatarsophalangeal joints by Latimer and Lovejoy (1990b) was qualitative, and the method behind it obscure. The former's more explicit measure of articular set instead suggested that this variable was indistinguishable between *Pan* and *Homo* and thereby devoid of information vis-à-vis arboreal-terrestrial or climbing-bipedal schema. Stern (2000) noted that dorsal and distal frontal (= coronal) perspectives serve to distinguish Great Ape and *Homo* first metatarsals, with humans displaying relatively flat and dorsally expanded metatarsal heads, respectively. The morphology of *Au. afarensis* aligns imperfectly within that of either apes or humans, again suggesting enhanced mobility and reduced load capacity of the first MTP joint with respect to *Homo*.

The Laetoli footprints would ostensibly be helpful for resolving the question of whether pedal anatomy indicates a completely committed terrestrial biped versus a merely competent biped with arboreal capabilities. To describe this literature as fraught is an understatement (see Tuttle 1985; Meldrum and Hilton, 2004), but the questions of hallucial abduction and the presence of a longitudinal arch are moot. Tuttle et al. (1991) offered that the apparent abduction of the toes in the footprints is an expected artifact of the naturally occurring "hallucal gap" of habitually unshod people. In addition, Berillon's (2004) analysis of osteological details of *Au. afarensis* pedal anatomy argued against the presence of a permanent medial longitudinal arch. The apparent arch betrayed in the Laetoli G-1 trackway is then a facultative one, induced by active muscular recruitment rather than by passive ligamentous and osseous reinforcement (Meldrum, 2004). These observations may be irrelevant for distinguishing facultative versus osseo-ligamentous arches, however, since there is evidence that active muscular recruitment is required for maintaining arch integrity in living people (Farris et al., 2020). Compliant and striding gaits do leave different signatures in footprints, however, and the trackways at Laetoli resemble the patterning of the latter more closely (Raichlen et al., 2010; Raichlen and Gordon, 2017).

Assessment of the hallucial anatomy of *Au. afarensis* supports neither an ape-like or human-like condition, but on the significance of this point, Latimer and Lovejoy (1990a) on the one hand, and Susman and Stern (1991) on the other, would most likely disagree. The discord stems from how traits that are "transparently" linked to bipedal progression should inform the more general question of locomotor behavior. The compromise camp concedes the presence and even the obligate nature of bipedality during terrestrial locomotion. Latimer and Lovejoy (1990a, 1990b) suggest that bipedality, once it was subjected to selection, required a complete abandonment of the trees, for a simple biomechanical reason: a foot adapted for bipedality is incompetent in climbing. With respect to the adducted (or incompletely abducted) hallux, "its absence is direct and virtually absolute confirmation that climbing behavior was negligible in *A. afarensis*" (Latimer and Lovejoy, 1990a:133). Their reasoning was that a foot compromised in climbing leads to an unacceptably high risk of falls (Latimer and Lovejoy, 1990b). The adducted hallux and short foot are incompatible with an arboreal habitus.

This combination counts as a magic trait when articulated in these terms: bipedal anatomy means that bipedality is all there is. Locomotor "ecology" reduces to selection for traumatic risk aversion (behaviorally) and efficiency (energetically). Several questions, however, may be raised. What, exactly, is the risk of falls in this case? All that can be said is that, with everything being equal, the risk is higher, *including* the assumption that the species in question must climb in a particular fashion. I expect some students of primate anatomy (but not field primatologists) experience initial disbelief when seeing goats and hyraxes in trees (Figure 4.2). After a certain interval, they can also witness these animals returning safely to the ground.

One problem with the above argument is in the units of comparison. Humans, on the one hand, and Great Apes, on the other, are both autapomorphic, relative to the morphology of *Ardipithecus* and its predecessors. These benchmarks are not ideal. Even if one counters that the locomotor milieus represent the best outcome selection can muster, the only reason to suspect that any species of *Australopithecus* is under similar selection pressures to humans or chimpanzees is to assume comparable ecological conditions.

The Innominate

Lucy (AL 288-1) preserves both an innominate and a sacrum, which permits reconstruction of the orientation of the ilium relative to cardinal planes. Because the innominate specimen was recovered in pieces, however, a precise reconstruction of iliac geometry is impossible. Stern and Susman (1983) provided a

Figure 4.2. Humans lack many of the anatomical features that facilitate climbing and other arboreal behaviors in nonhuman primates. This fact, however, should not be confused with an absence of a capacity to climb trees. These individuals of *Capra* lack grasping hands and feet and possess an intermembral index that suggests they are adapted primarily for terrestrial locomotion. Used under Creative Commons Attribution-ShareAlike 4.0 International License, user 923716947x, accessed at https://commons.wikimedia.org/wiki/File:Goat_Versus_Tree.jpg.

reconstruction that aligned the iliac blade close to a coronal plane, recalling a condition not unlike that of chimpanzees, albeit with a reduction in the height of the ilium. Lovejoy's (1979, 1988) reconstruction had the anterior iliac crest occupying a location intermediate between the sagittal and coronal planes, which is more convergent on the modern human condition (Figure 4.3). His reconstruction also suggested that Lucy's ilium had a more pronounced lateral flare than that seen in people today.

The combination of an iliac reorientation toward a sagittal plane, along with a more lateral iliac blade relative to the acetabulum, provides an effective lateral balance mechanism that is essential to energy-efficient bipedality. In Lovejoy's (1979, 1988) reconstruction, Lucy's lesser gluteals operated with an even greater mechanical advantage than those of modern humans. Under Stern and Susman's (1983) reconstruction, there was no opportunity for the lesser gluteals to function

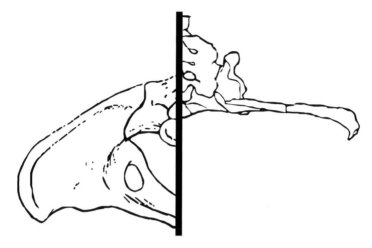

Figure 4.3. Two interpretations of innominate morphology in AL 288-1 ("Lucy") with respect to the orientation of the iliac blade. The major difference between Lovejoy's (1979, 1988) reconstruction (*left*) and that of Stern and Susman (1983, 1991) (*right*) is the degree and extent to which the lateral iliac blade is rotated into a parasagittal orientation. Dorsal is to the *top*. The superior perspectives in the reconstructions are not identical and are therefore not strictly comparable. The important difference between the reconstructions is nevertheless apparent. Although neither camp doubted that Lucy was bipedal, the mechanical implications of the two reconstructions are profound: under the compromise camp's reconstruction, there is little ability for the lesser gluteals to act as hip abductors, a critical muscular function in modern human bipedal progression. Lovejoy's (1979, 1988) reconstruction permits the claim that Lucy was at least as energetically efficient as modern humans with respect to hip-joint mechanics. Some aspect of this disagreement can be attributed to the condition of the fossil (Johanson et al., 1982); i.e., several cracks traverse the ilium, a certain distortion is evident in the auricular region, and some post-mortem rotation of the iliac blade relative to the acetabulum is evident. Both images are modified from Stern (2000).

as abductors to stabilize the trunk, as the posterior-facing iliac blade compelled an action of thigh extension for the lesser gluteals when the femur occupied a neutral ("mid-stance") position. The details of the contrasting reconstructions have obvious mechanical implications.

Apes, of course, are perfectly capable of bipedal progression, but they walk without full extension of knees or hips. They achieve lateral balance by recruiting the ipsilateral gluteus medius during stance phase; i.e., precisely as humans do. The difference in the action of the recruitment explains how. With a flexed hip, the gluteus medius in *Pan* medially rotates the thigh. With the support limb fixed momentarily on the substrate, this action amounts to lateral rotation of the trunk on the thigh, which provides a functional, if inefficient, mechanism of lateral balance (Stern and Susman, 1983). Here, electromyographic data offer an elegant explanation of biological role: the same muscle is recruited for lateral balance in both apes and humans, despite differences in muscle action and distinct postures during bipedality. This provides insight into hominin evolution by a mere demonstration of muscular activity. The transition to bipedality in this light is best viewed as an osteological rather than a neuromuscular transition. Lending further support for this idea is that the pattern of recruitment of the lesser gluteals is similar in climbing and walking, since, during the support phase of climbing, the gluteus medius acts as medial rotator of the thigh—crucial for maintaining support on vertical substrates.

In the absence of definitive evidence of the morphology and orientation of the ilium, the question of a fully human versus a transitional form of bipedality in *Au. afarensis* depends in large part on the evidence for and plausibility of a compliant gait in this taxon. The presence of genu valgus is not contentious, nor is the foreshortening of the superoinferior length of the ilium. On these criteria alone, the form of bipedality practiced by these early hominins was kinematically and energetically different from that of apes. But a compliant gait is energetically costly, in comparison with modern human bipedality, so the question becomes whether that form of locomotion would—or could—evolve. The position of Lovejoy et al. (1999, 2009) is that, from a developmental perspective, the options in terms of morphological alterations were finite. The problem currently is that we are not yet at the point where we can delimit those options. This is the unrealized promise of the structuralist paradigm.

Lovejoy's (1975) view prior to the description of *Au. afarensis* was that any form of transitional bipedality was inadmissible by the tenets of neo-Darwinian theory. This is a puzzling rationale, as such an orthodoxy concedes that despite the ubiquity of natural selection, other evolutionary forces, as well as pleiotropy and epistasis, conspire to keep populations from achieving phenotypic optima. A contemporary work defending neo-Darwinism (Charlesworth et al., 1982) also noted that the paradigm was pluralistic in terms of the details of phenotypic evolution. Indeed, Lovejoy's (1975) objection might be better framed through invocation of punctuated equilibria, but this theory emerged as hostile to neo-Darwinian sensibilities in the first place.

A more productive tack is to assert that compliant gaits ought to be selected against, given their thermodynamic inefficiency. This was the argument of Crompton et al. (1998): *Au. afarensis* would incur an intolerable heat load arising from the mechanical inefficiency of a compliant gait. As such, that locomotor pattern could not evolve. Stern (1999) offered a detailed rebuttal, arguing that this theoretically derived heat load was not supported empirically, and that the evolution of compliant gait did not necessarily assume human-like activity patterns (e.g., long-distance travel). Even so, no one disputes that a compliant gait is inefficient by a benchmark of human-striding bipedality. The absence of full knee and hip extension incurs unavoidable muscular work that modern people do not have to contend with, unless they are doing lunges at the gym.

Inefficiency is neither a guarantee of extinction nor an absolute impediment to evolution, however. When the benchmark of australopith bipedality is the modern human gait, then describing a compliant gait as ineffective or nonadaptive is misplaced. That benchmark is correct by the numbers, but it is contextually wrong. *Au. afarensis* was not competing with more-efficient bipeds (apologies to proponents of a deep-time ghost lineage of *Homo*), and, in any case, the proximate driver of success was foraging efficiency, accident avoidance (broadly defined), and inclusive fitness relative to conspecifics. Even if more efficient bipeds occupied the landscape, this does not logically compel extinction: only one species of antelope is the fastest, but we can see several antelope species running around at any one time. In chimpanzees, compliant bipedalism (obligate for them) is not demonstrably more energetically costly than conventional knuckle-walking (Pontzer at al., 2014; cf. Sockol et al., 2007). Even if this is not the case, added cost can be offset by changing foraging strategies. Most would agree that a foraging shift initially underlies the election to stand upright.

Perhaps *Au. afarensis* did, in fact, walk as we do. If so, then the problem of how—anatomically, metabolically, ecologically—bipedalism arose still exists, and in no way has it been solved. There must have been a phase of hominin evolution that involved a highly inefficient manner of bipedal progression. One can argue (without fossils, fruitlessly) about the duration of this phase in geological time, but only by invoking hopeful monsters could such a phase be avoided. Selection for bipedal energetics on par with living people would require that the combination of anatomical traits facilitating this would have been present in early hominin populations. These traits (genu valgus, low intermembral index, sagittally oriented anterior ilia, longitudinal and transverse plantar arches, etc.) did not appear simultaneously.

Lovejoy et al. (2009:73) expressed the view that *Ardipithecus* "is so rife with anatomical surprises that no one could have imagined it without direct fossil evidence." Arboreality was part of the locomotor repertoire of the genus, but this does not indicate a reconciliation of the efficiency and compromise camps. What it amounts to is an unstated concession that a kind of bipedality could coexist along-side a significant component of arboreality. But the anatomical details of *Ardipi-thecus* indicate that, just as the appropriate benchmark for early hominin biped-ality is not exclusively modern *Homo*, the model for early hominin arboreality should not be restricted to Great Apes.

Limb Proportions

As a group, australopiths were smaller than the vast majority of modern human populations in terms of stature and body mass. Both of these variables are esti-mated from skeletal remains using allometric equations. The fact that different reference samples for such equations yield disparate mass and stature estimates means that there is some degree of error in them. Nevertheless, there is little ar-gument that Lucy and her conspecifics were, by our standards, relatively small.

Debates over limb proportions in early hominins have revolved around ques-tions of the appropriate allometric corrections (see Jungers and Stern, 1983; Wolpoff, 1983). The relative lengths of forelimb and hindlimb have attracted the interest of functional morphologists because, among primates, they correlate with locomotion. For example, humans have a relatively low intermembral index, while African apes have high ones. This is explained in adaptive terms as relating to criti-cal aspects of locomotion. Long arms in orangutans are important for both sus-pensory behaviors and—in combination with a relatively short lower limb—for climbing. For modern humans, a long hindlimb translates into a longer stride length, which is energetically favorable and maximizes velocity. Theoretical sup-port for an association of climbing with a high intermembral index can be found in Cartmill's (1974) model, which linked competence in this activity to the produc-tion and maintenance of frictional forces in non-clawed mammals.

Lucy's preserved elements provided an opportunity to explore relative limb length in *Au. afarensis*. Under the empirically supported assumption that the cru-ral index (tibia length / femur length) does not distinguish humans and African apes, a humerofemoral index can be calculated as a surrogate for the intermem-bral index (Jungers and Stern, 1983). As direct comparison of values for taxa that differ dramatically in body size is unwarranted without an allometric correction, Jungers (1982) used a sample of Mbuti pygmies and a small bonobo to evaluate Lucy's humerofemoral index, in addition to recruiting catarrhine and hominoid predictive equations. He concluded that while Lucy's humerus was about as long as would be predicted for a modern human of her size, her femur was relatively short. Lucy's intermembral index has an intermediate value between those of humans and apes. This would mean, *ceteris paribus*, that Lucy would be better suited to climbing than modern people. But what this says about her general locomotor pattern is uncertain. The humerofemoral index of *Ardipithecus* is comparable to that of Lucy (85 vs. 89) and is well below that of African apes (Haile-Selassie et al., 2010). In noting that the australopith values were similar to those of *Proconsul*, Haile-Selassie et al. (2010:12125) described the limb proportions of the early hominins as "simply primitive." The modifier applied to the plesiomorphic status conveys the idea that there is nothing further to explain.

In fact, Haile-Selassie et al. (2010) dismissed the adaptive significance of the relatively short femur of *Au. afarensis*, via the assertion that because other critical modifications of bipedality had already arisen, the length of the hindlimb itself could not have been important as a target of selection. There are certain energetic disadvantages to having a relatively short lower limb, however, if bipedality is obligate in a terrestrial context. In the simplest terms, the easiest means to increase stride length (i.e., without major kinematic modification) in a striding or compliant biped is to increase limb length. For any given distance covered in a unit of time, fewer locomotor cycles entail less energetic cost. The relatively long foot in *Au. afarensis* may also have had energetic consequences. Susman and Demes (1994) used prosthetic footwear to substantiate the premise that angular excursions of the hip, knee, and ankle increase in proportion to foot length, and Foley et al. (1979) demonstrated greater vertical oscillations of proximal hindlimb segments in walking children, who happen to have relatively long feet, compared with adults.

These findings point to higher energetic costs, but the question is whether these costs were sufficiently high to require winnowing of short legs and long feet. Framed in the language of adaptation, the efficiency camp implies that selection was unnecessary *at that moment in time* in hominin evolution, while the compromise camp can be characterized as offering an alternative that can be expressed in two ways: (1) stabilizing selection against full commitment to either arboreality or terrestriality, or (2) directional selection against change toward the human condition, to maintain arboreal competence. In every case, there are assumptions about energetic costs that are, at best, incompletely quantified. But those intangibles pale in comparison with the unstated and generally unformulated assumptions about patterns of behavior.

The wrench, of course, is the environmental context. In Lovejoy's (1981, 2009) provisioning model of monogamous early hominins, it is hard to envision compliant gaits functioning very well, given a full load of savannah groceries added to the return trip. If bipedality involves frequent but short bouts of terrestrial foraging, then complaint gaits are inefficient, but to a trivial degree, considering relatively more time spent in the trees. A thermodynamic criterion of detecting adaptations is valid in the abstract, but it is not foolproof, because we can identify the evolution of traits and mistake this for insight into the selective regime that produced it (Leroi et al., 1994). Evolution is not necessarily adaptation, even if function is obvious (Hlusko, 2004).

The theoretical lacuna in the compromise camp was that utility could be assigned to every morphological condition. Thus everything was a candidate for adaptation. The efficiency camp made no such presumption. Instead, it claimed an ability to distinguish aptations from nonaptations, usually through character polarity. What has occurred in the debates over the origins of bipedality is that total morphological pattern or total biomechanical pattern are offered as guiding principles but—in the details of the argument—are abandoned in favor of a focus on traits as independent objects of selection. The structuralist sentiment is truly helpful once one gets down to specifics and can identify character combinations that are unbreakable due to developmental factors. Lovejoy et al. (1999) offer a schema of trait taxonomy that provides guidance about malleability and susceptibility to fixation by genetic mechanisms. This utilizes a framework for operationalizing a structuralist methodology, but the details of underlying process and diagnosis remain to be worked out.

What relative limb lengths mean in terms of biological role would seem to depend on whether the scientist asking the question is coming at it from a functionalist or structuralist bent. The default functionalist position is that a particular trait has been molded by activity. The primitive status of the trait is largely irrelevant. The structuralist perspective is that the trait is, first and foremost, understood to be inherited. Trait polarity, contextualized historically rather than formally (in a cladistic sense), potentially signals constraint. The expectation is that the functionalist and structuralist would agree on one point: the trait would be put to the best possible use, though they might easily disagree as to what that biological role would be, or if the trait in question qualified as a historical adaptation. This question of adaptive status is more of a problem for the morphologist than the organism.

In paleoanthropology, we have to believe that traits reflect behavior, or the endeavor is fruitless. This means that traits should be understood in a functional or mechanical milieu, but if they are selected and defined by their ease of quantification, then they must be contextualized after the fact. Lucy may have had an unremarkable humeral length by an allometric standard, but to argue that this negates competence in climbing represents a failure to contextualize, both in an anatomical and an ecological sense. In the former case, what matters is the limb length relative to the femur. In the latter instance, the utilitarian outlook of Susman et al. (1984) is convincing. No hominoid biped can hope to outrun a healthy and motivated hyena, irrespective of its energetic efficiency. Any ability to climb on short notice would be far more important.

The Bar-Glenoid Angle

Postcrania are relatively hard to come by in the hominin fossil record, and the scapula is a particularly rare element in the australopith inventory. Scapular morphometrics discriminate among primates fairly well (D. Green and Alemseged, 2004), but there is no consensus that the element possesses a strong functional signal (see Oxnard, 1967; A. Taylor and Slice, 2006; Bailey et al., 2017). The scapula plays no obvious critical role in bipedal locomotion, although the conventional wisdom is that the shoulder is freed from locomotor function, which, in turn, suggests that it can respond effectively to selection for novel, nonlocomotor functions.

The few scapulae of *Au. afarensis* are of interest, following the logic that this element should betray an arboreal habitus, provided the correct details of morphology are emphasized. Stern and Susman (1983) and Susman et al. (1984) calculated a bar-glenoid angle for AL 288-1,[5] which, they argued, separated Lucy from modern human samples. By contrast, this variable was more similar to modern apes (within one standard deviation of *Pongo* and *Gorilla*). As this angle is intended to reflect the orientation of the glenoid, the ape-like angle in Lucy signifies a more cranial orientation. Hence, "this trait was an adaptation to use of the upper limb in elevated positions as would be common during climbing behavior" (Stern and Susman, 1983:284).

The acknowledged fly in this ointment is that Lucy was undoubtedly smaller than any of the members of the human or gorilla comparative samples.[6] Some allometric consideration has to be given to ascertain what the angle value would be in diminutive gorillas or a blown-up australopith. Susman et al. (1984) anticipated this objection, extrapolated the human sample to Lucy's size, and recalculated the predicted human value; the resultant confidence interval excluded Lucy. Inouye and Shea (1997) avoided the problem of extrapolating beyond sample ranges by employing ontogenetic series of Great Apes and humans. They found an absence of scaling in the apes but a relatively noisy signal for the human sample. Predicted angles for humans and apes converged in the lower size range of the samples, approximately where Lucy was situated. The implication of these data is that there is no morphometric basis for differentiating Lucy from humans, once the size "correction" has been applied. Stern (2000) conceded this possibility but raised the question of whether ontogenetic series represented an appropriate test for what was originally formulated as a static adult functional comparison.

It is noteworthy that the numbers are what is left to tell the story of locomotor reconstruction. Once again, the functional calculus is that if "fossil equals ape," then "fossil moves as the ape does." The bar-glenoid angle has functional significance ostensibly because it betrays glenoid orientation, but in terms of the actual argument, AL 288-1 was arboreal because its measured value is unlike that of humans. What is lacking is due consideration of what the consequence of differences in mean angle (on the order of 15 degrees) has on function *in vivo*. Humans maintain the considerable shoulder mobility of apes, so shoulder elevation, by itself, seems unlikely to be what the adaptation is. Larson and Stern (1996) and D. Green and Alemseged (2014) suggested that cranial orientation enhanced joint stability during suspension and climbing. This implicates relative muscle size and lines of action as part of glenohumeral stabilization. Larson's (2015) survey of scapular myology, however, indicated that the inference of muscle size and activity from indices of scapular form (i.e., relative size of the infraspinous fossa) was unreliable.

Even with additional *Au. afarensis* scapulae being recovered in recent years, the significance of glenoid orientation is still contested (see D. Green and Alemseged, 2004; Haile-Selassie et al., 2010). Some of these arguments are over measurement error, but they also stem from an adherence to the idea that morphometric separation along taxonomic lines provides an unambiguous window into function. Haile-Selassie et al. (2010:12123) inferred that "comparison of KSD-VP-1/1g with other hominoid scapulas suggests functional uniqueness." This statement was largely based on comparisons of several angular measures of the glenoid, ventral bar, scapular spine, and vertebral and axillary borders, and the diagnosed uniqueness relies on the distinctive position of the fossil relative to the compulsory comparative samples of humans, chimps, and gorillas for each variable. In no case was the fossil shown to be significantly distinct from all of the modern species. It is fair to ask on what basis a functional differentiation was justified, as only a proximate morphometric one has been demonstrated. The performance attributes— or even functional capacity—of the observed disparities were undeveloped.

The KSD-VP-1/1g scapula is nearly complete and, as such, provides more information relative to the single angle extracted from AL 288-1. Haile-Selassie et al. (2010) consequently invoked total morphological pattern as a justification for an argument of the unique functional morphology of the specimen. The authors did not specify what functions were involved, beyond stating that AL 288-1 may have been "post-arboreal," or have evolved from an undefined unique locomotor pattern, or both. One appeal to the total pattern was an oblique reference to the modular evolution of the scapula outlined by Young (2004). The insinuation is that modularity is functionally driven and presumably reflects the action of selection on the modules in question, but Young (2004) was explicitly agnostic on this point.

The idea that patterns of morphological integration provide more direct insight into function is intriguing, but it is not particularly compelling. The identification of integrated traits provides circumstantial evidence of functional coordination or cohesion, but common function is only one possible agent of developmental integration. The hypothesis of integrated development of fore- and hindlimb serial homologues (Hallgrímsson et al., 2002), for example, would seem to throw up a barrier to functional or historical adaptation, given the different biological roles, kinematics, and loads experienced in primate fore- and hindlimbs. Modularity presumably permits more-direct and efficient action of natural selection on an anatomical region, under the logic that the functional traits within are integrated, but this observation provides no unique insight into specific functions, only guidance into questions of evolvability (e.g., Cheverud, 1982; Wagner and Altenberg, 1996; Hlusko et al., 2004; Villmoare, 2013; Parins-Fukuchi, 2020).

The Femoral Neck

The skeleton, given its principal functional status as the structural scaffolding of vertebrates, is often analogized to engineering structures. The femoral neck was one of the inspirations for Wolff's Law. Paleoanthropology has embraced this analogy of the femoral neck as a cantilever in the search for unambiguous signals of human-like bipedality. Stern and Susman (1991) identified cortical bone distribution of the femoral neck as a "magic trait" employed by Lovejoy (1988).

If the femoral neck is loaded as a cantilever, then there is tensile stress acting longitudinally on the superior margin of the neck and compressive stress is present along the inferior margin. The action of the lesser gluteals in stance phase (with perhaps horizontal-component contributions from some other muscles) superposes an axial compressive force. If these stress components are summed, then the net stress on the superior femoral neck is very low, and the compressive stress along the inferior border is absolutely higher. Under this model, the critical region of bone reinforcement is along this inferior border (Figure 4.4). Lovejoy (1988) and Ohman et al. (1997) examined the femoral neck of humans and *Au. afarensis*, observing thin cortical shells along the superior femoral neck, with

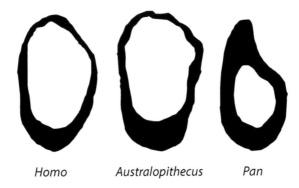

Homo   Australopithecus   Pan

Figure 4.4. Cortical geometry of the femoral neck in (*left to right*) humans, *Australopithecus afarensis*, and chimpanzees. Humans and the fossil are more similar to one another with respect to a thickened cortical shell along the inferior aspect (*bottom of images*), a postulated result of the superposition of axial and bending loads encountered in bipedal progression. This comparison is standard in paleoanthropological research, with the default logic being that the resemblance of fossils to humans represents human-like capacity, and the resemblance to apes represents ape-like capacity. In this case, it is also assumed that bipedalism, being a rare form of locomotion, also generates a relatively uncommon morphology. This turns out not to be true once a wider survey of primates is undertaken (see Lovejoy, 1988; Stern and Susman, 1991; Ohman et al., 1997; Rafferty, 1998; Demes et al., 2000). Modified from Stern (2000).

thicker bone inferiorly. Since their survey of non-hominin hominoid femoral necks revealed a more equitable distribution of cortical bone around the neck, the comparison of apes, humans, and fossils "demonstrates unequivocally that habitual terrestrial bipedality was the primary locomotor behavior of *A. afarensis*" (Ohman et al., 1997:130).

The dichotomous nature of the observed pattern has been challenged along several fronts. Stern and Susman (1991) provided images of a bonobo proximal femur that exhibited the human "terrestrial biped" pattern. Rafferty (1998) also observed the "biped" cortical distribution pattern in most of the 20 nonhuman primates she examined, and Demes et al. (2000) found thin superior cortical bone along the superior femoral neck to be the norm among strepsirhines.

This trait—cortical asymmetry in the femoral neck—is appealing, because the expectation is that cortical thickness will faithfully track variation in load magnitude. The functionality of the trait is easy to understand: bone will reinforce where the net loads are highest.[7] Therefore, cortical bone is assigned the useful property of being a reliable monitor of load severity on a local scale. This hijacks the trajectorial theory of trabecular architecture to explain the idiosyncrasies of cortical reinforcement. If the premise is valid, few problems of biomechanical inference are beyond the grasp of the paleoanthropologist. Unfortunately, this is not true (Demes et al., 1998, 2001).

One way to try to salvage this "unique" bipedal signal is to dismiss all comparisons that stray outside the Hominoidea as inappropriate, under the banner of a narrow phylogenetic bracket. This would not be terribly convincing for a number of reasons, including the need for a convenient, arbitrary, and unsubstantiated assumption that the skeleton's response to load type and magnitude is qualitatively different across primate higher taxa. Certainly bone's metabolic response to the mechanical environment depends on genetic variation to some degree (I. Wallace et al., 2012), but without a demonstration to that effect, the operative assumption should be that bone—as a tissue—has certain physiological responses to mechanical stimuli that are, on the whole, applicable across vertebrates. In theory, it would be easy to refute this assumption through targeted investigations, but across-species comparisons under identical experimental designs are relatively rare.

Another unexamined assumption in this particular debate is the degree to which the load cases are qualitatively distinct. There is very little direct empirical evidence to support theoretical estimates of joint load magnitude and direction under a variety of postures, irrespective of species-specific anatomy. In climbing or bipedality, the orientation of the bending moment is in flux throughout stance phase. The cantilever load, as presented, is—in the best-case scenario—a gross oversimplification.

Closer attention to engineering theory suggests that the cantilever is actually the wrong model. The extensive literature on long bone mechanics has enabled a more or less unconscious default to beam theory as an explanatory stricture for the primate skeleton. Even if the load case, as presented by Lovejoy (1988) and Ohman et al. (1997), is generally correct, the presumption that bending is responsible for the critical stresses may not be. Demes et al. (2000) remarked on the abbreviated femoral necks seen in some strepsirhines and suggested that the more immediate danger in this region is shear. This is a valid alternative, because classical beam theory considers normal tensile and compressive stresses to be most important, relative to shear stress. For this idea to hold, the standard engineering textbook admonition is that if the span:depth ratio of the beam is greater than 10:1, the shear stresses are comparatively small and can be ignored. Femoral necks are nowhere near this threshold, and shear stresses represent a significant fraction of normal stresses, in terms of magnitude. Another consideration is that while bone is famously weak in tension, under shear it is weaker still. Thus it is likely that the critical component of stress in the femoral neck, from a structural integrity per-spective, is shear. The fact that Demes et al. (2000) found no relationship between femoral neck length or angle and cortical asymmetry provided circumstantial sup-port for the idea that beam theory is a poor choice for explaining cortical geom-etry in the femoral neck. Finally, the high density of trabecular networks in this region is consistent with a loading environment imposing significant amounts of shear. The above considerations fit what Skedros (2012) deemed the "shear-resistance priority hypothesis," in which an efficient strategy for ensuring ade-quate bone metabolic activity is to have the osteocyte network be most responsive to shear strains.

Given that shear is the most important component of stress acting on the fem-oral neck, the functional explanations offered for cortical geometry in this region are no longer relevant. The reason is that, under shear, variation in cortical geom-etry is largely inconsequential. The amount of bone, in contrast to the particu-lars of its arrangement, is the primary determinant of structural integrity. Another source of shear stress that is seldom considered with respect to the femoral neck is torsion. For this to occur, the resultant load must not be coincident with the centroid of the femoral neck. How likely that is depends on one's point of view. If the femoral head can be loaded eccentrically, and the vertical component of the abductor resultant is not aligned with the centroidal axis, then some amount of torsion will occur. On the other hand, Latimer et al. (1987) argued that nonor-thogonal loads cannot occur between complementary articular surfaces in syno-vial joints, owing to diminutive coefficients of friction. In any case, a superposi-tion of torsion and axial compression of the femoral neck is as satisfactorily explained by the details of cortical geometry as it is by the bending and compres-sion surmised by Ohman et al. (1997).

Exchanges in the literature regarding the significance of femoral neck cortical geometry in *Au. afarensis* amount to a dog's breakfast. The load case is assumed to be monolithic for a biped, and to be a variety of unspecified loads for anything else. Cortical geometry is explained under an inappropriate model. Despite data permitting calculation of areal section properties for estimating relative stresses, neither Ohman et al. (1997) nor Stern and Susman (1991) provide any ballpark fig-ures on what level of stresses or strains would have been involved, which is al-ways a good means to check on the general validity of a mechanical model. Con-sequently, understanding the significance of proximal femoral morphology requires consideration of the effects of body size, as this is highly relevant to load magnitude. Obviously, the variable for determining mass needs to be something other than features on the proximal femur. The demonstration that femoral head size is relatively small in *Au. afarensis* (Jungers, 1988a, 1988b) suggests that joint loads at the hip were relatively reduced in comparison with modern humans, irrespective of femoral neck geometry. This is compatible with the idea that early hominins utilized compliant gaits.

Who Does One Believe?

The debate over locomotion in *Au. afarensis* surrounds two points: (1) whether competent arboreality was performed by *Au. afarensis* with any regularity, and (2) whether the bipedality practiced by this species differed kinematically or kinetically from that of modern humans. Beyond the morphometric analyses that are central to specific functional inferences, there have been frequent appeals as to why a particular view is more credible. These rely on specifying a selective imperative that, if unfulfilled in the Pliocene, would have meant certain extinction. Tree climbing was central to the compromise camp, since, without it, Lucy and her ilk faced unsustainable rates of predation. In this view, the only performance threshold that mattered was utter incompetence in shimmying up the tree, since even "bad" climbers would still remain alive enough of the time. The view of the efficiency camp was that the transition to bipedality necessarily compromised arboreal performance, as the cost of suboptimal performance was exceedingly high. Latimer and Lovejoy (1989, 1990b) contended that increased probabilities of falls from trees would spell certain doom, since even single falls have severely deleterious effects on fitness. The data marshaled for their argument are silent with respect to these points, even though the chain of logic—linking functional competence to variance in fitness—is exactly what ecological morphology should aspire to. In the paleontological context, getting at the relevant ecological variables under specific behavioral scenarios is daunting, and there are questions that should inspire skepticism with respect to both. Does competence in arboreal behavior require an anatomical vicar for African apes? Are trees the only option for protection from predators? Do primates that fall from trees have zero fitness? The answer to all three is no. Many species not named *Pan* or *Gorilla* can climb trees. Primates do fall from trees and break bones with nontrivial frequency, but healed fractures indicate that these events are survivable (Schultz, 1939; Jurmain, 1997; Nakai, 2003). The idea that African ape–level competence in climbing is prerequisite to arboreality is simply false.

The observation that the two camps have inherent biases is true, but this characterizes every scientific endeavor. Any theoretical framework is selective with

respect to what facts and variables are important. More helpful for one's understanding is to ask whether there are anchors of data or theory that predetermine
which conclusions will be drawn, irrespective of the form of subsequent data. An
anchor is simply an object of attention (among many potential ones) that predis-
poses an inference. As Kahneman (2011) notes, human decisions are heavily biased
by anchors, and their inferential validity can be poor. They are utilized because
they are available. Here I am invoking this idea with considerably more license
than its use in Kahneman's (2011) sense, which is restricted to a quantifiable value
that influences subsequent estimates. But the efficiency camp, for understandable
phylogenetic reasons, had a concept of arboreality that was anchored on an African ape design criterion. The idea that other anatomies could negotiate an arboreal environment was largely ignored until *Ardipithecus* forced the issue.

The underlying assumptions of the two competing research programs are
largely unstated but, when articulated, were often used for rhetorical effect rather
than epistemological commitment. The declared adherence to neo-Darwinian
principles by the efficiency camp is vacuous, because this can be contorted to
mean anything with respect to interpretation of specific variables. Beyond the
structuralist or functionalist leanings of the contending positions, ideas of how
trait lability should inform evolutionary scenarios were not explored. Little more
than the proximity of the fossils to a confidence interval was used to advance a
position. The underlying assumptions behind variable choice were often unexamined at the outset. For instance, the presence of curved (i.e., inhuman) phalanges
was originally presented as evidence for arboreal behaviors (Stern and Susman,
1983; Susman et al., 1984), but supportive evidence only emerged much later (Richmond, 2007; Rein, 2011; Congdon, 2012). Similarly, potentially validating data
from *in vivo* investigations (Berillon et al., 2010) postdated the reconstructions of
metatarsophalangeal joints in the fossils. These examples underscore that resolution requires a painstaking examination of the degree to which (1) linear or angular variables reflect biomechanical performance (e.g., in terms of joint excursions or stress), (2) actual behavior approaches the inferred performance limits,
and (3) a thermodynamic criterion can be connected to inclusive fitness in some
tangible (nonarbitrary) way. Analytical and experimental techniques exist to address all three of these concerns.

Lest the reader surmise that I have punted on offering an opinion as to the ve-
racity of arguments in this historical debate, I will close this chapter by restating
what I consider to be the pivotal issues concerning these locomotor reconstructions. First, the ecological context is absolutely critical. We cannot recruit the

morphology of the fossils themselves to piece together a landscape to which the organisms in question were adapting. In doing so, we only reassure ourselves that we are omniscient. If we are to be credible functional morphologists, we need independent environmental reconstructions. Fortunately, those exist (e.g., Bedaso et al., 2013; Behrensmeyer and Reed, 2013). The question is whether these reconstructions are sufficiently fine grained to allow alternative functional hypotheses to be critically evaluated.

The dimorphism in size among the Hadar fossils is also accompanied by differences in morphology that are argued to have locomotor consequences (Stern and Susman, 1983; Senut and Tardieu, 1985). If there is a single hominin species represented at Hadar, this raises the possibility that the large morph (presumably males) and the small morph (presumably females) have somewhat distinct but overlapping locomotor repertoires. While this idea has been met with varying degrees of skepticism (Stern, 2000; Lague, 2002; Ward, 2002), in principle there is no reason why locomotor preferences and capacity cannot vary within a taxon. Most of us would be surprised were this not the case. With variation comes uncertainty, however, and from the perspective of rhetorical persuasion, it is usually unwanted.

The details of arguments in this historical debate are, for the most part, exercises in functionalist thinking. The claim that primitive characters are adaptively uninformative is disguised as a retreat from functionalist orthodoxy, but this is a case of cherry picking from the structuralist toolbox for purposes of excluding swaths of data. Functionalism relies on the idea that historical adaptation is the only arbiter of evolutionary change, and everything else is just noise. This is logically indefensible, because it provides no role for behavioral plasticity in negotiating the environment, nor does it allow that traits can be exapted for novel biological roles while still maintaining their morphology. A structuralist perspective is the correct one for interpreting morphology, but the incorporation of ideas on constraint have yet to make a substantive impact on actual analyses of morphological and biomechanical variables in paleoanthropology.

Structuralism undermines the ideal of total morphological pattern as the guiding principle of functional inference. Under structuralist logic, there must be morphological characters that are relatively immune to selective forces that may heavily influence the evolution of other characters. This should not serve as an endorsement of magic traits. As noted before, the term was introduced by Stern and Susman (1991) as a warning against overinterpretation of anatomical details. Ohman et al. (1997) rose to the bait to declare that there were no such things, and they invoked total biomechanical pattern as the only way to begin to understand locomotor behavior. Without a hint of irony, they declared that the cortical bone of the femoral neck provided an unequivocal demonstration of habitual terrestrial bipedality.

Far too much faith is placed in the largely unconscious notion that statistical differentiation indicates proportionate divergences in functional performance or biomechanical capacity. With respect to individual variables, there are two under-explored questions with respect to morphological traits. How was it used? And what energetic and performance consequences follow from observed variations? Seeking the answers from direct observation of fossils is impossible, so we are left to speculate. This necessary realm of supposition is where we have important choices to make, analytically and methodologically.

Chapter Five

Hominin Dietary Adaptations

Hominins became differentiated by an ecological shift of some kind. As paleoanthropologists, we are to figure out this new ecological context from paleoenvironmental clues, some of which will be based on fossils of bones and teeth. Explanations of bipedality with respect to fitness, as reviewed in the previous chapter, ultimately revert to the question, What is this mode of locomotion good for? Foraging is central to all the competing hypotheses; even Lovejoy's (1981) model, which emphasized interbirth intervals and pair-bonding, had provisioning as a critical component of the adaptive complex. Since the details of foraging are a matter of diet, it follows that identifying what early hominins ate might go a long way toward understanding why they stood up in the first place and how we humans ended up dominating the biosphere.

With respect to early hominins, the adaptive significance of various details of skull morphology—in terms of function—remains contested and unresolved, even though we seem to have less difficulty making taxonomic decisions with this material. Partially explaining these disagreements is the fact that there are few good analogs for the hominin feeding apparatus, especially with respect to the australopiths. There are simply no living primates or mammals that resemble the fossil crania of these early hominins very closely.

As discussed in Chapter 3, however, this has not deterred publication of arguments for inferring hominin dietary adaptations through analogy. Those analogies are necessarily forced, because the anatomical dissimilarities of the taxa under comparison beg the question, and are testable primarily through counterexample. Morphology can be sidestepped in inferring the primordial hominin diet using paleoenvironmental criteria, but eventually the morphology has to be reconciled with the foods being exploited in those landscapes. Stable isotope and dental microwear analyses provide independent tests of dietary hypotheses emerging from morphological study, and, in this sense, the inference of diet is more straightforward than particulars of locomotion.

A number of features of the dentition and craniofacial skeleton characterize early hominins. These include postcanine megadontia, reduced canines, anteriorly positioned zygomatic roots, variable degrees of sagittal and nuchal cresting, and expanded mandibular corpus dimensions (Figure 5.1). These features find their

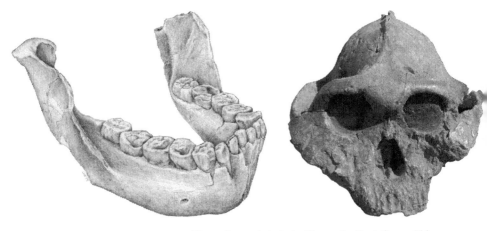

Figure 5.1. Two iconic specimens of *Paranthropus boisei*, the Natron (or Peninj) mandible (*left*) and KNM-ER 406 (*right*). The facial skeleton characteristic of this species can be attributed to some unique aspect of feeding behavior and diet along two lines of reasoning: its configuration indicates an ability to produce relatively large occlusal forces (Rak, 1983; Eng et al., 2013), and its relative size indicates a structural strategy for maintaining adequate safety factors for mitigating the destructive effects of either large and/or repetitive masticatory loads. The sagittal crest and large, flaring zygomatic arches bounding an exceptionally large temporal fossa are considered to be indicative of a massive temporalis muscle. Similarly, the depth and thickness of the zygomatic arch itself, as well as the tall and wide ramus of the Natron mandible, are suggestive of a large masseter muscle. These chewing muscles would have been effective for producing very large biting forces, and/or they may represent muscles that are particularly fatigue resistant. In either case, bony reinforcement of the facial skeleton is structurally required for maintaining bone integrity. The buttressing of the facial skeleton is most pronounced in terms of the massive anterior root of the zygomatic arch and mandibular corpus dimensions. Natron jaw illustration by Luci Betti in Daegling and Grine (2017), reprinted by permission from Springer Nature; KNM-ER 406 image by Yiselaat, accessed at https://commons .wikimedia.org/wiki/File:Paranthropus_Boisei_-_Transparent_Background.png.

greatest expression in the "hyper-robust" *Paranthropus boisei*, but their status as derived characters is already discernible in the earlier hominin forms, *Australopithecus anamensis* and *Au. afarensis*. Whether trait expression is regarded as one end of a continuum or as truly autapomorphic leads to two competing explanations for *P. boisei* morphology: (1) it represents the extreme expression of an initial hominin dietary adaptation, or (2) a major shift in food resources explains a unique morphological configuration. Whatever the case, the reduction in postcanine size in early *Homo* probably signals either a decreasing reliance on the teeth for initial food preparation or reflects a fundamental dietary shift away from the australopith condition.

The function of teeth is to break down food, but this observation, by itself, does not elucidate biological roles at the species level with adequate precision, beyond the insight that the metabolic cost of comminuting peanuts in the oral cavity is less than the proximate energetic savings of swallowing them whole.[1] The details of occlusal morphology, however, are widely regarded as yielding information about the mechanical properties of ingested and masticated foods (e.g., Kay, 1975; Boyer, 2008). Allometric considerations are integral to such analyses, and I will explore why this is so presently.

Postcanine Megadontia

One outstanding feature of australopiths is that they had very big teeth, especially in the postcanine battery, and this applies whether the benchmark of comparison is canine size, skull size, or body mass. The postcanine megadontia of *Paranthropus* species naturally invited allometric arguments as an explanation. Pilbeam and Gould's (1974) foray was appealing in the sense that there was nothing to explain, other than that each taxon was meeting metabolic requirements linked to body size.[2] This argument is not convincing for at least two reasons. First, as noted in Chapter 2, interspecific differences in australopith body size are fairly modest (Jungers, 1988a; McHenry, 1991, 1992). Second, adult chimpanzees are close living analogs in size, and their jaws and teeth are relatively diminutive. In *Paranthropus*, we see the culmination of the australopith trend in molarization of the premolars, progressive enlargement of the molar battery, and a marked reduction in canines and incisors. Reduction of the anterior dentition is typically explained in one of two ways: (1) they were small because they were generally unimportant for processing the diet, or (2) they were small because there was no room for large front teeth once selection focused on making the postcanine ones bigger. These are not mutually exclusive explanations, but both view the diminutive teeth of *Paranthropus* as functionally impoverished. This recalls Aristotle's principle of excess and defect, in that both function and structure are seen as a zero-sum game. The large cheek teeth exact a metabolic or somatic cost that is paid for via a reduced anterior dentition.

Paleoanthropologists have looked upon the phenomenon of postcanine megadontia almost exclusively in terms of adaptive explanation. The alternative—that drift or pleiotropy play a role—is seldom seriously entertained, because, by an energetic criterion, the large investment in tooth volume would be relatively costly. Under this logic, baboons could benefit from postcanine megadontia, but the socioecological benefit of canine displays and the use of incisors for processing a variety of foods preclude this. The Aristotelian view is that there is a finite amount of morphological "stuff" available developmentally, a nebulous and unhelpful principle unless redefined in more explicit energetic terms. An alternative explanation is that postcanine megadontia creates functional compromises in a baboon facial gestalt that are intolerable. This idea should be taken seriously if the cost of compromise is laid out in terms of both performance and structural integrity.

There are at least three needs that could explain postcanine hypertrophy. The first is that an expanded occlusal area offers a platform for bulk processing. This explanation features in most allometric arguments, but it does not necessarily hinge on scaling. Any diet in which calories or nutrients are in short supply per unit of ingested food is costly, and one mitigation strategy is to increase food throughput. Whether added occlusal area decreases the daily workload is no doubt dependent on the mechanical particulars of diet. Large point loads on small teeth get the job done just as well as those on big ones, but distributed loads that do not need to be large could be advantageous with bigger teeth.

A second reason for large teeth is maintenance of structural integrity: teeth will break if the stresses inside them get high enough. The cost of breaking a tooth is obvious, and it involves some finite loss of food-processing rate. Selection for structural integrity in diphyodont taxa, such as primates, is undoubtedly strong. Models of tooth chipping and fracture suggest that thickened enamel limits fracture risk (Lucas et al., 2008; J. Lee et al., 2011; Constantino et al., 2012). These ideas, as articulated, require occlusal forces to be large or teeth to be small for selection to be targeting crown integrity, because stress is the critical variable. Bite forces have been inferred to be unusually high in the australopiths under these models, but bite pressures (stresses) were not unusual in these taxa, in comparison with living apes (Demes and Creel, 1988; Eng et al., 2013).

A third advantage of large teeth is durability of the occlusal surface, not with respect to whole-tooth fracture, but instead to the accumulation of microscale fractures at the interface of teeth with each other and with intervening food particles. This wear process degrades function gradually, rather than instantaneously. The lifespan of a functional tooth crown is extended if, to begin with, there is more material to be removed. What is particularly remarkable about *Paranthropus* fossils is that their molars have very thick enamel caps, and parts of these are completely worn away in some adult specimens. This observation suggests that the enamel cap's function of resisting fracture has an accessory benefit: promoting tooth longevity with respect to abrasion and attrition. One can also simply invert the argument: selection for wear resistance has an accessory benefit of fracture resistance. The built-to-not-break argument is preferred, given that the cost of tooth fracture is presumably high, relative to the more incremental costs of attrition. Undoubtedly, wearing teeth out produces a loss of efficiency in terms of food processing that, superficially, would seem to be similar to breaking a crown. Yet the performance loss for single teeth may not be very critical. Dental senescence is compensated for by behavioral adjustments in living primates (Millette et al., 2009), such that loss of function through wear is not a death sentence.

Enamel chipping, which is observed with some frequency in australopith teeth, represents an intermediate state between attrition and whole crown fracture. One reason why the dental chipping evidence led to inferences of hard-object feeding as the principal dietary strategy of australopiths is that thick enamel in living primates has been linked to durophagy (Hylander, 1979; Kay, 1981; McGraw et al., 2012). Rabenold and Pearson (2011) examined how enamel thickness covaried with phytolith load in primates and found that abrasion potential explained as much of the variance in enamel thickness as the consumption of hard seeds. Pampush et al. (2013) argued that the link between enamel thickness and diet is indirect, since enamel thickness is associated with expected longevity in primates, irrespective of diet. These observations indicate that the adaptive significance of enamel thickness in australopiths is not resolved, even if all the aforementioned functional benefits are transparent.

Occlusal Morphology and Bunodonty

Beyond the functional ramifications of variations in tooth size and its effects on wear and structural integrity, consideration of occlusal topography has figured prominently in discussions of australopith diets. Compared with that of African

apes, australopith dentition is bunodont. Whether the association of limited occlusal relief with thick enamel is compulsory is incompletely understood, but it
could be attributed to one of two (or both) functional requirements: breaking food
down effectively, or extending the working life of a tooth. From either a performance or maintenance perspective, the occlusal features of australopiths, as a
group, would appear to be congruent with a diet involving hard seeds as an important component (Kay, 1981; C. Peters, 1987; Butler, 2009). Conversely, the absence of sharp-shearing crests on these teeth is inconsistent with a folivorous diet,
per the logic that two-dimensional foods are not shredded efficiently by blunt
teeth. Selenodont ruminants and lophodont primates serve to reinforce this view,
and high occlusal relief is associated with folivory in primates (Kay, 1975). While
it is indisputable that the shape of teeth and mechanical properties of diets (ag-
gregated coarsely) are correlated (Boyer, 2008), this general line of argument is
more or less purely utilitarian. The large *Paranthropus* molars, with their thick
enamel, inspired the paradigm of a nut-cracker. Strait et al. (2013) and A. Smith
et al. (2015) argued that since the australopith facial skeleton evolved adaptations
for feeding on hard foods, the bunodont cheek teeth should be interpreted in the
same light. That is, these teeth, being functionally ill-suited for shredding vege-
tation, could not be considered adaptations for such a diet. The contention is un-
abashedly functionalist: once the ecological source of selective pressure is identified, all related morphology in the functional complex becomes targeted to that
singular adaptive end.

Given a primate comparative framework, the *Paranthropus* dentition is decid-
edly suboptimal for a herbivorous diet, if we are referring to leaves, grasses, and
sedges.[3] Suboptimal does not mean nonfunctional, however, or even ineffective.
With respect to any aspect of biomechanical performance, the determinant of adaptation is functionality relative to conspecifics, other ecological competitors, or
minimal life-sustaining performance at crucial junctures of life history. Efficacy,
in terms of optima or esthetics, is irrelevant. Merely bad performance in a pool of
really terrible performers can yield excellent fitness. There are many ways to be
good enough, and, with respect to primate dentition, the emphasis has been on
the geometry of occlusal surfaces, or the overall topography. In comparative surveys, researchers typically avoid exploring this relief on worn teeth (e.g., Ramdarshan et al., 2011), recognizing that wear changed the functionality of the original
(and implicitly "adaptive") geometry. Ungar and M'Kirera (2003) and Berthaume
(2014) identified contexts in which wear itself was adaptive by increasing the number and extent of sharp working surfaces at exposed enamel-dentin interfaces.

The application of methods not dependent on homologous occlusal features, such as Dirichlet normal energy (Bunn et al., 2011) and relief index (Ungar and Williamson, 2000), capture functionality of whole occlusal surfaces. Berthaume et al. (2018) applied such methods to examine the "sharpness" of australopith teeth compared with those of Great Apes. The perspective of this approach is that fissures, crenulations, cusps, and exposed edges of worn enamel constitute a set of shearing features on an occlusal surface. While the gross relief seen on chimpanzee and gorilla molars reveals an intuitively sharper tooth by australopith standards, topographic analysis comes to quite a different conclusion: there are more comminution tools on the australopith teeth. This is counterintuitive, until one appreciates that this conclusion primarily reflects a shift in the scale of the analysis. In effect, australopith dentition is well-suited for producing the type of shearing forces needed to process tough foods. Berthaume et al. (2010) also examined a sample of australopith tooth models to examine the effects of variation in occlusal morphology on the efficacy of hard-object fracture. That study failed to support the idea that different topographies were better at producing brittle fracture. The authors instead suggested that selection had acted to maintain structural integrity of the teeth under the loading environments associated with durophagy.

According to this scenario, in which the australopith postcanine dentition is large, comparatively indestructible, and effective for processing various foods, the answer to the query, "What did *Paranthropus* eat?" could logically be, "Anything they wanted to." This does not settle the question of what the feeding adaptation was, nor does it constitute a refutation of the nut-cracker paradigm per se. It does fit within Wood and Strait's (2004) framework that *Paranthropus* could be described as an ecological generalist, a fossil hominin example of Liem's paradox.

Ideally, the examination of disparate (and preferably independent) lines of evidence yields consilience as far as understanding selective pressures induced by diet. In terms of anatomical sources of inference, the dentition is regarded as the most direct line of evidence, but the correlation between occlusal relief and diet masks a large asymmetry of precision. The characterization of occlusal morphology may be excellent, but diet is usually defined categorically, without taking variance within diets into account.

Facial Skeleton

In tying diet to the functional significance of various features of the skull, relative size is everything. As is the case with the dentition, the logic is not hard to follow. For example, considering mandibular corpus size, the large jaws of australopiths could reflect (1) large body size, (2) differences in tooth support requirements, or (3) a biomechanical response to unusually large masticatory forces (see Figures 2.5 and 2.6). An examination of each possibility leads to subtle differences in interpretations of dietary adaptation.

If somatic size suffices to explain the size of the mandible, then the crucial consideration is how diet covaries with body size. Among mammals, dietary quality declines as body size increases (Eisenberg 1981); this applies to primates as well (Kay 1975). This reflects an allometric principle based on metabolism: small animals have relatively high metabolism, gut transit times are fast, and low-quality foods that are consumed in bulk are useless because what nutrients there are can't be effectively extracted. By contrast, larger animals can afford to ingest junk, because their metabolism is relatively slower, and they can take extra time to process nutrients. Applying these principles to hominins, East African *Paranthropus* had to eat more in absolute terms, but less relatively (as a proportion of body mass), than its australopith brethren (Pilbeam and Gould, 1974). Buried in this argument once again, however, is the expectation that tooth size is tightly linked to body size. When Pilbeam and Gould (1974: 900) warned, "We cannot judge adaptation until we separate such changes into those required by increasing size and those serving as special adaptations to changing environments," they were stating that an allometric explanation could supplant an ecological one. There need be no philosophical objection to the idea that—as expected allometric variants on some ecological theme—australopiths represent a "grade" of hominin evolution,[4] regardless of the details of their phylogeny. In the end, the allometric argument is obviated, since the assumed differences in body size among australopiths have not survived scrutiny. The taxonomic differences in the mandibles are due to something other than body size.

Mandible size could simply reflect the simple functional need for the bone of the jaw to support bigger teeth, or to develop those teeth within the mandible during growth. Both Dart (1948) and Wolpoff (1975) suggested that jaw proportions in australopiths were simply effects of tooth size: the jaws were big because the teeth were. Though seldom articulated explicitly, what this might indicate is that the extraordinary biomechanical properties, which confer high bending and torsional strength in australopith mandibles, are merely artifacts of selection for growing and supporting a large dentition. R. Smith (1983) suggested that mandibular robusticity (traditionally conceptualized as breadth/height of the corpus) in australopiths was readily explained by the reduced canines of the radiation, on the grounds that large canine roots compelled deeper mandibles.

Comparative study does not support the assertion that mandibular size in early hominins is simply an expected outcome of growing and supporting big teeth (Daegling and Grine, 1991; Plavcan and Daegling, 2006; Figure 5.2). Australopith jaws are apparently larger than they need to be, either for developing teeth or for structurally supporting their erupted dentitions (Figure 5.3). From the energetic criterion of adaptation, large jaws should be fulfilling some function to offset the costs of their growth and maintenance. An alternative is that australopith jaws came to their morphology as an effect of something else (either drift or pleiotropy), but the favored explanation is some sort of mechanical adaptation. There are a number of ways in which this can be investigated, but it is not obvious what aspect of mechanical performance should be focused on.

The upper and lower jaws would seem to be reciprocal, mechanically speaking, since they house complementary sets of teeth and have the same functional end: preparing food for digestion. Despite a great deal of research into the biomechanics

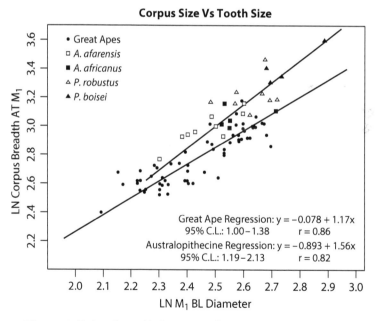

Figure 5.2. The atypically broad mandibular corpus of australopiths has been interpreted as an effect of postcanine megadontia (Wolpoff, 1975), the logic being that large tooth crowns necessitate large tooth roots, and that these, in turn, must be supported by large mandibles. A regression of tooth breadth on corpus breadth suggests that this is an incomplete explanation. For a given tooth size, the mandibular dimensions of the fossil hominins are larger than expected. Daegling and Grine (1991), used with permission from John Wiley and Sons.

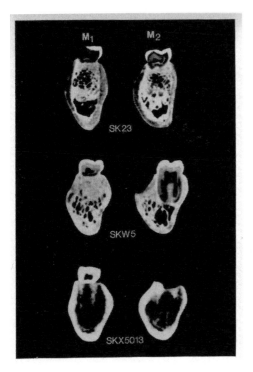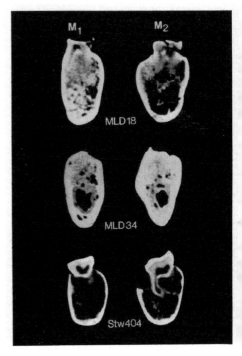

Figure 5.3. CT scans of mandibular cross-sections of *Pararanthropus robustus* (*left*) and *Australopithecus africanus* (*right*). Daegling and Grine (1991), used with permission from John Wiley and Sons.

of the facial skeleton, both as a whole and in parts, there are few agreed-upon dietary signals in the primate face. The upper facial skeleton has several functions, in addition to stress resistance and dental support. In particular, the sensory capsules for olfaction and vision are found here. Moss (Moss and Young, 1960; Moss, 1968, 1973) argued that the soft tissues and spaces of these sensory capsules are primary determinants of local skeletal architecture. Even if this is not strictly true, there are certainly functional compromises that need to be accommodated throughout the facial skeleton. The mandible, as well, must permit competing functions of respiration, vocalization, ingestion, and mastication that dim hopes of unobstructed insight into feeding behavior by virtue of its morphology. Mandibles, however, have been the focus of the majority of studies that have proposed linkages of skeletal morphology to diet.[5]

Australopith mandibles are exceptional in terms of both size and geometry, with *Paranthropus* jaws being superlative. Whether scaled relative to tooth size

(Daegling and Grine, 1991), jaw length (Hylander, 1988), or inferred body size (McHenry, 1991), the cross-sectional dimensions of australopith mandibles are larger than expected, given the patterns seen in living or other extinct primates. Before Hylander's (1979) experimental model of the sources of masticatory load in primate mandibles, the peculiarity of australopith mandibles was not described in terms of explicit biomechanical properties. Generally they were contrasted with early *Homo* jaws by their "robusticity," without further comment (e.g., Robinson, 1954; Wolpoff, 1975; Day, 1977).

Hylander (1979) borrowed concepts from engineering's strength-of-materials subdiscipline to describe mandibular function in terms of stress and strain—or, more generally, in relation to strength and rigidity. This perspective treated the mandible as a load-bearing structure, rather than as an effective food processor, but, most importantly, it was based on experimental strains that were caused by actual rather than hypothetical loads. As a result, these data constrained future arguments about masticatory forces.

A focus on load-bearing capacity presents no particular interpretive issues with respect to feeding, since the strength of the jaw faithfully reflects how much force it can produce: a strong mandible is undoubtedly also an effective food masher. With the mandible conceptualized as both a beam and a twisted member, a largely mysterious loading environment became tangible, and jaw function could be treated numerically, permitting desired contrasts to be evaluated in a more objective fashion. While the argument was meticulous, it was also simple: jaws being bent should be deep, and jaws that were twisted should be broad. This model suggested that primate hard-object feeders and folivores could be sorted out from standard-issue frugivores, based on mandibular geometry, once size effects were evaluated (see R. Smith, 1983; Bouvier, 1986; Ravosa, 1996, 2000). Hylander (1988) noted that australopith mandibles, in particular those of *Paranthropus*, were both deep and broad, relative to modern primates. Under the axiom that form follows function, this was interpreted in most quarters as indicating something unique about australopith feeding and mastication, with *Paranthropus boisei* being the paramount example.

Australopith jaws were relatively strong. On the business end of the teeth, this means that bite forces could be very high, and the jaws would be no worse off.[6] This would be an ideal state of affairs for eating seeds and cracking nuts. Hylander (1979) originally noted that large mandibles would also be beneficial for resisting fatigue failure. If australopith mandibles were selected to mitigate fatigue risk, this coheres to a dietary regime that involved bulk processing of low-quality foods, which is consilient with one of the benefits of megadontia (Groves and Napier, 1969; Swindler and Sirianni, 1976). Thus, whether or not australopith mandibles were adapted to be structurally strong, there are two disparate dietary specializations that could explain their evolution. The first is a reliance on hard objects, such as nuts or seeds, requiring high masticatory forces. The second is a dependence on low-grade forage, requiring a great deal of chewing. Moreover, one of these explanations does not preclude the other (J. Wallace, 1975).

There have also been other dietary hypotheses. The idea that carnivory was the behavior separating hominins from the apes is probably very old (Darwin obliquely suggested this in *The Descent of Man*), with Dart (1948) formally arguing the case for *Australopithecus* at Makapansgat. Neither functional morphology nor biomechanics had substantive influence on this or subsequent hypotheses of carnivory. This is probably due, in part, to the lack of models among living primates.[7] Termites may have been sought after by *Paranthropus robustus* (Backwell and d'Errico, 2001), but fossil anatomy had nothing to do with this idea, either. Meat-eating and insectivory are both reasonable strategies for early hominins and may represent critical resources in some contexts. Such behaviors would appear to be irretrievable from morphological details, but we cannot reject this possibility on the basis that their faces resemble neither hyenas nor anteaters.

Thus all but two hypotheses are ostensibly resolved through biomechanical or functional analyses of the facial skeleton: (1) australopiths, as a group, became increasingly specialized for hard-object feeding, with *Paranthropus boisei* being the supreme nut-cracker; or (2) australopiths were "herbivorous" in general, with *Paranthropus* having to exploit that particular diet more intensively. In the biomechanical literature, durophagy has received more attention. This could be because it is the better hypothesis, but it also may be an effect of the law of the hammer. Loading a hard sphere to failure is a fairly simple biomechanical problem to solve, once the model conditions are worked out. In contemplating a model of an amorphous herbivorous diet, figuring out what these conditions should be is more daunting.

There are two explanations why durophagy is the default hypothesis on australopith feeding adaptations: the first is ostensibly evidence-based ("empirical"), while the second is epistemological (in terms of the historical precedence of certain ideas). This contrast is a simplification, because the particulars of an idea condition what observations will be sought in supporting it. But the juxtaposition of these sources of adaptive inference is useful for understanding their influence on functional arguments.

It is incorrect to argue that the moniker bestowed on OH 5 ("Nutcracker Man") anchored ideas on the preferred feeding adaptation of *Paranthropus boisei*. Leakey used the term in a 1958 article appearing in the *Illustrated London News*, but in his subsequent announcement in *Nature* there was no mention of such behavior. Instead, he opined that OH 5 had been feasting on flesh from the bones of other species found at the site, with the help of stone tools (L. Leakey, 1959). The popular analogy took hold, however, and it made sense under a purely utilitarian criterion of function: big flat teeth were good at not being cracked themselves while they broke open the nuts lodged between them. The adaptation was obvious, but anthropologists were not altogether uncritical. Living ecological analogies were offered both in support of (Jolly, 1970) and in challenge to (Cachel, 1974; Szalay, 1975) the concept of seed-eating as the central dietary adaptation. These analogies was heavily influenced by the ideas Dart (1948) had expounded for *Australopithecus prometheus* (see Grine, 1981)—in particular, the premise that hunting was the behavioral solution for the hominin invasion of terrestrial, open-country habitats. In one sense, in the second half of the 20th century, the question of australopith feeding adaptations was often interpreted in contrast to (and more rarely in agreement with) the "man the hunter" paradigm that grew out of Dart's (1948) propositions.

Robinson's (1954) dietary hypothesis—that *Australopithecus* and *Paranthropus* morphology indicated omnivorous versus herbivorous tendencies—was revisited by Du Brul (1977), who invoked a bear analogy to posit that *Australopithecus* is to *Paranthropus* as *Ursus* is to *Ailuropoda*. Analogies invoked outside of primates are rightly met with increased skepticism,[8] but Du Brul's work was explicitly biomechanical, as opposed to ecological. With respect to bite force production, temporomandibular joint stability, and occlusal morphology, his paper offered a persuasive case that *Paranthropus* was highly derived and probably a dietary specialist, relative to the "gracile" australopiths. He did not settle on a particular diet for *Paranthropus*, other than to say it was "herbivorous," which left an appropriate amount of wiggle room.

Roughly contemporaneous with the announcement of *Australopithecus afarensis*, there were a variety of hypotheses on *Paranthropus* diet competing for attention in the paleoanthropological community: carnivory (Dart, 1948; Cachel, 1975; Szalay, 1975), herbivory (Robinson, 1954; Du Brul, 1977), and graminivory (Jolly, 1970). Durophagy featured prominently in two of these hypotheses, with Szalay (1975) presuming that megadontia permitted bone-crushing, and Jolly (1970) favoring seed-eating. Rak's (1983) elegant, qualitative biomechanical analysis of australopith facial evolution largely steered clear of dietary inference, although a few illustrations indicated that he saw the primary adaptation as one for crushing large seeds or nuts.

The discovery and interpretation of *Australopithecus afarensis* reframed the dichotomy of "gracile" versus "robust" australopiths, because *Au. africanus* was no longer the stem australopith. Instead, this taxon was now seen as derived specifically toward *Paranthropus*. Rak (1983), in particular, viewed *Au. africanus–P. robustus–P. boisei* as progressively specialized in their facial skeletons. This, in effect, diluted the perceived ecological differences between South African australopiths, at least in terms of how morphology was thought to reflect feeding adaptations.

While a utilitarian paradigm views the australopith dentition as not just suited for, but adapted to hard-object feeding (Strait et al., 2013), an important question is whether craniofacial morphology leads directly to the conclusion that *Paranthropus* (or *Au. africanus*) was adapted in that manner. Two finite element models argued that this was the case (Strait et al., 2008, 2009 for Sts 5; A. Smith et al., 2015 for OH 5). In each case, the inference hinged on the idea that the facial skeleton would be prone to potentially dangerous stresses were the buttressing features of the midface not present. The adaptation, in effect, was the avoidance of a particular level of stress in a certain region of the face under a specific load case. Provided these components were all modeled correctly, the biomechanics of the situation were accurately reproduced. The important question, however, is whether selection acted to bring about this state of affairs.

The first issue is whether the explanation is congruent with the mechanobiology of bone. The skeletal biology literature establishes that there are many paths by which to get skeletal hypertrophy,[9] such as a moderate number of really high loads every day (e.g., cracking nuts), or a large number of moderate loads daily (e.g., eating leaves). Both of these options can have more or less the same effect on bone mass. As noted in Chapter 2, the interaction of load magnitude and frequency boils down to strain rate. Any value of strain rate can be obtained by a theoretically infinite number of combinations of its constituent variables. If this is how the strain environment is interpreted on a cellular level, then (from the viewpoint of feeding) the identity of the signal does not imply an identity of function or performance.

The potential impact of phenotypic plasticity looms large in this context. Empirically, there is strong support for the idea that ontogenetic shifts in dietary or feeding regimes induce plastic responses, which contribute to adult variance in facial bone material and structural properties (Ravosa et al., 2007, 2016; J. Scott et al., 2014b). In most discussions, *Paranthropus* morphology is characterized as monolithic, so, by virtue of this practice, the role of plasticity can be completely ignored. An essentialist portrayal of morphology treats variance as somehow epiphenomenal to function. J. Scott et al. (2014b) employed a rabbit model to demonstrate that seasonal versus constant feeding on mechanically challenging food conferred different plastic responses with respect to bone mass in the mandible and palate. A hay diet required significantly more daily masticatory work to process and was associated with increases in bone mass, relative to controls. Most provocative, however, was that for intermittent cessation of hay feeding (the "seasonal" contrast), the differences in bone mass vis-à-vis controls eventually diminished. J. Scott et al. (2104a) interpreted this as an experimental refutation of the "fallback"—or seasonal—exploitation of hard foods championed by Strait et al. (2009). Simply stated, hypertrophy of the jaws can be directly linked to the consistency and duration of chewing bouts on a daily basis. Intermittent high loads (e.g., nut-cracking) have no special explanatory power with respect to skull "robusticity."

Changes in skeletal traits can be due to both directional selection of alleles and simultaneous alteration of plastic responses (Garland and Kelly, 2006). Thus the question of whether plasticity or intrinsic "constitutive" change (i.e., in which morphological change is linked to allelic substitutions) is not an either/or proposition. The two processes can operate concurrently to produce "fit" individuals in response to environmental change. At the same time, the existence of plastic reaction norms is no guarantee of individual success: features regarded as "pathological" may be exemplars of phenotypic plasticity (Ravosa and Kane, 2017). The buffering of phenotypes is insurance against deleterious plastic responses, reinforcing the view that plasticity has its costs.

The mechanobiological perspective, applied to primate jaws, is currently agnostic on the question of dietary signals in skeletal variation. This is not a limitation of the viewpoint per se, but rather an acknowledgment that there is substantial variation in experimental outcomes. These are attributable to different model species, varying populations sampled within species, and noncomparable interventions.

Since it would be unrealistic to assume that accounting for sampling choices and experimental design would eradicate residual variance in such studies, there are other reasons to be skeptical. As C. F. Ross et al. (2012) detailed, if diet is tangibly linked to craniofacial morphology in primates, there have to be predictable associations among intervening factors of food properties, feeding behavior, loads imposed on the bones, and the resultant stress and strain fields. Such correlations will be imperfect, and the noise between factors is compounded as one moves from diet to morphology, or vice versa. In addition, it is worth considering that there may be a "file-drawer" problem with comparative research attempting to link facial morphology to diet; i.e., negative results tend to be locked away and do not have much of a presence in the primary literature. This is hard to establish, as the confirmatory data are inaccessible, but it is reasonable to suppose that, in the literature, there is bias for meaningful correlations to be reported, whereas investigators will be less likely to pursue publication of ambiguous or insignificant comparisons, and reviewers tend to regard such data as uninformative. Likewise, if support for a preferred hypothesis is the goal of a research effort, the omission of "noncooperative" taxa can be easily rationalized.

A final reason to question a specific biomechanical link between the australopith face and durophagous adaptation is comparative data. There are several hard-object specialists among living primates (pithecines, *Sapajus* capuchins, and mangabeys) and none of them resemble *Paranthropus* in terms of orthognathy or postcanine megadontia. Though there are certain parallels (thick enamel, molarization of P4 in *Cercocebus* mangabeys), the prognathic faces of living hard-object specialists suggest that the australopith face is sufficient, but not necessary, for durophagy. In this case, morphology does not provide an unambiguous window into diet. Other data are needed.

Nonmorphological Means of Inference

Two lines of evidence have been particularly useful for inferring diet in the paleoanthropological record, and they are, to different extents, independent of morphology. This is useful for adaptive inference, since these approaches offer insights into diet that do not depend on an understanding of the biological role of a morphological character. Stable isotope analysis samples bone and dental tissue and provides information on the chemistry of consumed foods during a finite period. It does not reveal diet per se, but it can be used to circumscribe the types of food resources utilized during an animal's life. Similarly, dental microwear analysis does not permit the inference of specific foods ingested, but it is assumed to reflect something about the mechanical properties of consumed items. Both techniques have been applied extensively to the australopith fossil record. Most paleoanthropologists see these methods as a means of validating functional or adaptive hypotheses.

Microwear analysis in paleoanthropology initially recruited a small but convincing data set from living mammals to argue that distinctive wear signatures on occlusal surfaces are linked to ingested foods (A. Walker et al., 1978; Teaford and Walker, 1984). Grine (1981, 1986) examined molar microwear from *Au. afri-canus* and *P. robustus*, arguing for a more generalized diet in the former and a pref-erence for nuts and seeds in the latter. This provided early validation for the nut-cracker paradigm: *Paranthropus* molars were heavily pitted and displayed what would now be described as complex yet isotropic wear fabrics (R. Scott et al., 2005). Grine's (1981, 1986) studies and other contemporaneous work also focused on the reconstruction of occlusal movements from wear signatures, an endeavor that is conspicuously absent today. These early efforts concentrated on measuring the dimensions of individual microwear features, a painstaking and, when sam-pling among populations of features, an arguably subjective approach. More-automated image analyses, such as Fourier transforms (Grine and Kay, 1988) and dental microwear texture analysis (R. Scott et al., 2005), appeared to validate the distinctions in wear between the "gracile" and "robust" forms. The association be-tween complexity of microwear and hard-seed ingestion has also been observed in contemporary primates, but the diagnosis of diet from these studies remains coarse, with highly folivorous versus durophagous diets contrasting most clearly in microwear fabrics.

The idea of an australopith grade of human evolution lends itself to an infer-ence of a singular—even if broad—dietary adaptation, which Strait et al. (2013) argue was hard-object feeding of varying types and degrees. The South African australopiths offer agreement with this idea, as all three species are plausible seed-eaters by a microwear criterion. The East African australopiths, however, display microwear fabrics that are more reminiscent of modern primate folivores and *The-ropithecus* than of other hard-object specialists (Ungar et al., 2008, 2010).

Predictably, two resolutions are offered regarding this disparity between cra-niofacial anatomy and the record of ingested foods impressed on the teeth. The first is that the preferred interpretation of anatomy is wrong, and the second is that microwear does not record feeding behavior accurately. Microwear does have se-rious limitations as an indicator of diet. One is that turnover of microwear fea-tures on enamel is very rapid—on the order of days or even after only one or a few occlusal loads.[10] Another is that materials other than foods are effective at scor-ing and breaking enamel. In defending the nut-cracking adaptation of *Paranthro-pus boisei*, Lucas et al. (2013, 2014) argued that foods themselves are largely inca-pable of producing microwear, with the real culprits being environmental grit and dust. In fact, the observation that mineral matter contributed to wear had been substantiated decades before (Kay and Covert, 1983), but, since the comparative data from living primates succeeded in differentiating broad dietary differences, it had been assumed that grit loads in primate foods did not swamp a dietary signal. More recent work substantiates this assumption (Burgman et al., 2016; Adams et al., 2020; Hua et al., 2020).

In the context of australopiths, the declaration that microwear is bogus as a dietary indicator relies on some questionable premises. The first is that South African dirt and dust is qualitatively distinct from that in East Africa (Strait et al., 2013), a claim that Ungar et al.'s (2016) data suggested is without merit in the context of microwear formation.[11] In addition, high microwear feature turnovers mean that rarely ingested but critical foods—perhaps those to which a species is adapted—will be invisible in a probabilistic sense. This invocation of fallback foods—critical resources that may not be a frequent or preferred item—permits microwear to be ignored completely as a line of evidence (Strait et al., 2009; cf. Grine et al., 2010). This reasoning has the undesirable property of rendering an untestable hypothesis into an invisible piece of decisive evidence.

Arguments over microwear's value continue to rage, because, for the most part, the technique's validity is correlational rather than processual. The etiology of feature formation is difficult to model theoretically, and experimental demonstrations are de rigueur, but none sufficiently re-create the dental and oral environments. The meaning of the patterns can be debated, in large part because the underlying process of feature formation remains poorly understood.

Stable isotopes sampled from dental tissue offer a different window into diet. Isotopes accumulated in dental enamel provide a longer-term record of diet than can be supplied by microwear studies. In addition, since the timing of dental development is fairly regular, isotopic data have built-in temporal controls, furnishing a glimpse of dietary variation over early stages of the lifespan. Ratios of ^{13}C to ^{12}C in sampled tissues reflect proportions of ingested plants that utilize different photosynthetic pathways. C_3 plants include trees, bushes, shrubs, and some grasses, while C_4 plants comprise warmer-climate grasses and many sedges. A reasonable, but not infallible, generalization is that C_3 foods will be found and consumed in forested environments, while C_4 plants will be associated with more open-country habitats. For example, arboreal folivores would have a high C_3 isotopic signal, in contrast to *Theropithecus*. Nut-crackers are expected to cluster with species having C_3-dominated diets.[12]

Australopith enamel has been extensively sampled in East and South Africa. In East Africa, *Au. anamensis* isotopic values are similar to those of African apes and indicate a predominantly C_3 diet (Cerling et al., 2013). This finding is congruent with morphology, given that, among australopiths, *Au. anamensis* is least derived from *Pan. Au. afarensis*, by contrast, was more heavily reliant on C_4 resources, and this diet was apparently relatively stable though geological time, despite inter-individual variation (Wynn et al. 2013). The *Paranthropus* species (*P. aethiopicus* and *P. boisei*) both show a preference for C_4 resources (Cerling et al., 2011, 2013). It is an understatement to describe these findings as unexpected. From function-alist corners, this was met with the rejoinder that independent measures of diet constituted poor ancillary evidence of dietary adaptation (Strait et al., 2013).[13]

The isotopic evidence from South Africa suggests a diverse diet, with seasonal and inter-individual variation, in *Paranthropus robustus*, with some shifts between C_3 and C_4 resource exploitation (Sponheimer et al., 2006). The interpretation of these data was that *P. robustus* was better modeled as a dietary generalist than a specialist, with a foraging strategy more akin to modern baboons than chimpan-zees. In light of Liem's paradox (see Chapter 3), this interpretation is reasonable, and, because hard-object feeding remains on the table, it is also uncontroversial.

Isotopic analysis of *Au. africanus* teeth yielded a similar mix of C_3 versus C_4 di-ets. (Sponheimer and Lee-Thorp, 1999). The exploitation of C_4 resources, in con-trast to modern chimpanzees, indicates greater dietary breadth and, probably (like *P. robustus*), exploitation of different habitats over a lifetime or seasonally. The later *Au. sediba* has isotope ratios indicating an "almost exclusive" C_3 diet (Henry et al., 2012). Notwithstanding the small sample (two individuals), the eschewing of C_4 resources is intriguing, if for no other reason than the craniofacial anatomy of *Au. sediba* is, in some features, more reminiscent of early *Homo* than *Australo-pithecus* or *Paranthropus* (see de Ruiter et al., 2013; Kimbel and Rak, 2017).

It is important to note that the isotopic signatures do not invariably reflect in-gestion of plants. Animals and insects that are consumed will bring their own isotopic signatures of whatever plant matter they happened to be munching be-fore they themselves became food. Eating a grazer will up the C_4 signal, while in-gesting a worm that feeds on forest saplings will do the same for the C_3 fraction. Meat consumption, then, may be indicated in the isotopic data, but there is no current means to parse out the plant versus animal signal. A reluctance to specu-late is thus analytically prudent, although it is worth considering whether there is a Dartian hangover in play. Revisiting the possibility of an osteodontokeratic culture is unpalatable, although we can invoke a very broad (*Papio*) or narrow (*Pan*) phylogenetic bracket and conclude that opportunistic scavenging or deliberate hunting are reasonable foraging behaviors for early hominins. Sponheimer and Lee-Thorp (1999) suggested that animal consumption is at least as reasonable as plant-based sources for explaining the C_4 signal in *Au. africanus*.

Table 5.1   Approaches to dietary reconstruction

Method	Directness of inference	Specificity of inference
Mechanical models of feeding	low	low
Dental microwear analysis	high	medium
Dental topographic analysis	low	low
Stable isotope analysis	high	low
Phytolith analysis	high	high

Note: "Inference" refers to the efficacy of various methods for reconstruction of diet. Dental topographic analysis, for example, maps onto the details of diet only in general terms, and it has relatively low discrimination for what a given individual may have eaten. Phytoliths sampled from fossil teeth, however, permit a direct inference of ingested food (owing to its association with the fossil) and specific dietary information (by taxonomic identification of the ingested item).

To summarize, the isotopic data indicate that the earliest australopiths and *Au. sediba* exploited dietary niches of a breadth comparable to that of chimpanzees; *Au. africanus* and *P. robustus* expanded their dietary repertoire commensurate with more-diverse habitat exploitation; and *Au. afarensis* and *Paranthropus* in East Africa relied most heavily on C_4 resources (grasses, sedges, and succulents). The microwear data indicate eclectic feeding habits in South African forms, while the East African australopiths display low-complexity microwear, which is generally at odds with known microwear associations with hard-object feeding.

An even more direct means of inferring ingested foods is the analysis of phytoliths adherent to fossil teeth. Assuming that the association of phytoliths and teeth is not a postmortem taphonomic artifact, this would seem to provide an unambiguous window into diet, given the taxonomic specificity of the recovered phytoliths (Ciochon et al., 1990; Henry et al., 2012). Even though phytolith–tooth associations will be fortuitous, they will be effective for evaluating competing dietary inferences (Table 5.1).

Reconciling Contradictions

If the dietary and anatomical data are interpreted as paradoxical, then one of the sources of inference is misleading, or possibly both. The apparent disjunction emerges because the nut-cracking paradigm is foundational. Without this prior assumption, there is no reason to contemplate the "impossibility" that *P. boisei* was adapted for exploiting C_4 resources. The historical context is nowhere more relevant than in this case. Rak (1983) conceptualized the australopith face as a trend of facial retraction and reinforcement along an *Au. africanus–P. robustus–P. boisei* axis, with the implication that this was a progressive enhancement of a singular dietary adaptation. Cognitive dissonance arises when the adaptation to hard-object specialization is switched out for another one: a focus on grasses, sedges, reeds, and the like.

There should be no rational objection to the possibility that an evolving lineage, or sister taxon, can undergo shifts in feeding ecology and yet recognizably remain members of their clade. Dietary flexibility in living primates warns against using morphology as a simple metric for circumscribing diet (see Brown and Zunino, 1990 for tufted capuchins; Codron et al., 2006 for chacma baboons; McGraw et al., 2007 for guenons). *Paranthropus robustus* fits the paradigm of the nut-cracker, and this species' anatomy, microwear, isotopes, and available habitats are consilient. *P. boisei* might fit the paradigm in terms of habitat and anatomy, but nut-cracking is not supported, other than from a prior commitment to the reasoning that its anatomical configuration is only possible through a single selective pathway toward exploitation of a specific type of dietary resource.

The observation that flat, large molars are not ideal for shredding foliage may be true, but in the evolutionary context, the important question to ask is, Where are you coming from? In the case of *P. boisei*, you evolved from a mess kit that includes thick enamel, relatively flat teeth with numerous small topographic features, and a set of muscles with a supporting skeleton that enable you to chew powerfully and often. Whoever your immediate ancestor is, you do not have the option of turning down this inheritance and reconfiguring everything.[14]

A scenario can be envisioned in which, whatever the dietary repertoire of East African *Paranthropus* was, it became less available, perhaps due to competition with contemporaneous *Homo* or environmental changes affecting habitat availability. For the purpose of argument, assume seeds and nuts were the primary resources that became scarce. What else could be exploited? If selection could continue along the same trajectory—enlarging occlusal areas, thickening enamel, maintaining a variety of occlusal "tools," buttressing the skeleton, and growing big masticatory muscles—anything that could be ingested could be processed. A lack of efficiency would only matter if the caloric investment in chewing precluded the diversion of somatic energy to reproduction. Maintenance of the same anatomical configuration is perfectly plausible in the mechanobiological context. The shift away from hard foods to tough ones would still impose a stress stimulus on the facial skeleton that would encourage bone maintenance.

Scenarios such as this may be proof of nothing other than the imaginations of their authors. The foregoing constitutes no meaningful test of the hypothesis that feeding on C_4 plants is the underlying adaptation for the craniodental morphology of *P. boisei*. What the scenario does accomplish is to provide a speculative foundation from which to explore additional possibilities for feeding behavior. This may be one case in which identification of adaptation is not the most engrossing aspect of paleobiology: biological roles may be more important for understanding *P. boisei* as a species. The partitioning of biological role from adaptation is instructive in the case where one identifies the *P. boisei* face as exapted for grasses but adapted for hard-object feeding.

Biomechanics is an essential component of functional morphology, but the peril of applying it via the paradigm method is that it elevates the biomechanical model to the arbiter of adaptation. These models should not have that status. They can describe what is possible, what is efficient, and what is strong, but this does not predict what natural selection will do, nor does it specify the accompanying behavior. The argument offered by Strait et al. (2009, 2010) more or less stated that finite element analysis solves the question of australopith feeding adaptations which microwear and isotopes cannot. Their decisive test was identification of critical mechanical threats, with the variables of strain energy and stress magnitude as the crucial metrics. In effect, their contention was that a particular load case (premolar loading in isolation), in which the threat to skeletal integrity was greatest, was the principal driver of facial morphology. Preventing damage and failure was the biomechanical adaptation. (The efficacy of the system with respect to feeding was a secondary consideration and was incompletely explored.)

While Strait et al. (2009, 2010) argued that consilience with ecological and other data was realized, there are several problems in having the biomechanical model be the primary means of adaptive inference.[15] Most alarming (or optimistic, depending on one's point of view) is the belief that the modeling of only two load cases—in a masticatory apparatus capable of a diversity of mechanical behaviors—provides an unambiguous window into feeding ecology. It is unlikely that direct insight into feeding behavior follows from such an analysis. What the model does provide is some clue as to what happened when Sts 5 bit on something large with its premolars. That is valuable to know, but the adaptive inference is reckless. Reeve and Sherman's (1993) call for an evaluation of phenotype sets is nowhere more relevant than here.

Despite the failure of FEA in solving problems of adaptation in the fossil record, finite element modeling can be productively applied to mechanobiological problems (Carter and Beaupré, 2007). These recursive models make two important contributions to understanding skeletal morphology. The first is that, to a large degree, skeletal morphogenesis can be understood as mechanically mediated. The second is that the initial conditions of the model (which are usually unknown and therefore approximated) have large effects on model behavior. A stress analysis, in isolation, describes mechanical behavior but offers no insight into the process by which that stress field came to be. The finite element models applied in paleontology are not incorporating mechanobiology into their assessment of evolutionary adaptation. Instead, the mechanical criterion is purely utilitarian, echoing the paradigm method insidiously through a more sophisticated analysis. Important questions are unanswered. Why would demonstration that premolar biting incurs potentially damaging stresses mean bone has responded to that isolated mechanical context? The fact that the bone is potentially compromised suggests instead that the adaptation is ineffective. This observation is anathema to the paradigm, but it is coherent with respect to the mechanobiological perspective. This is explored further in Chapter 6.

A Productive Role for Contingency

Understanding morphogenetic processes—the proximate manifestations of phylogenetic constraint—will surely improve our comprehension of morphological changes in the paleontological context. Our knowledge about such processes in fossils will necessarily be limited, and this will probably make us less certain of some of our adaptive inferences. We would do well to embrace that uncertainty in reconstructing human evolution.

Diet is not the same thing as feeding behavior, and the latter is what must be understood or reconstructed to understand "dietary" adaptation. With hominins, we are obviously dealing with primates who have the cognitive capacity to circumvent mechanical problems posed by selected foods. Humans cook. Chimpanzees use rock hammers and anvils to smash nuts. There is no reason to think that australopiths did not also have the capacity to solve food-processing problems before ingestion.

Dart and Craig (1959) envisioned australopiths as homicidal maniacs, a behavior inextricably linked to their carnivorous habits, but two very different analyses prompted a new reckoning. Brain (1981) applied the nascent science of taphonomy to demonstrate that Dart was probably quite wrong. Jolly (1970) suggested that seed-eating made the hominins what they were. The value of both of these works was not that either was right; this is ultimately unknowable. Their contribution was that they inspired and incubated new lines of research. Both were imaginative in different ways. They were top-down ecological studies that took existing knowledge and reapplied it to new sets of assumptions. Importantly, both are testable.

A Structuralist Perspective on the Early Hominid Skull

A review of arguments over dietary or feeding adaptations reveals that interpretation of craniofacial morphology in the hominin fossil record has been almost exclusively functionalist in nature. The messy fit between the ancillary data and morphological variation at the species level should invite some skepticism as to the suitability of the functionalist worldview in this instance. Paleoanthropologists have recruited comparative data from living primates to link morphology to function, but that fit has been anything but consistent (Daegling and McGraw, 2001; C. F. Ross et al. 2012; McGraw and Daegling, 2020). The unstated assumption is that the goodness of fit in selected cases, considered together with the axiom that bone is a finely tuned, environmentally sensitive tissue, justify the functionalist logic.

From the standpoint of evolutionary biology, a structuralist perspective is the correct one, but it does not make adaptive inference easier. The idea that mandibles correlate with dietary variables obviously does not demonstrate that the mechanical impact of these variables molded mandibular morphology. Two examples illustrate this point. The first is that the species-specific attributes of the mandible are generally present well before weaning, and feeding variation during ontogeny does little to turn these mandibles into anything recognizably different. This does not call for a total retreat from a mechanical explanation for form generation, but only for a recognition that epigenetic factors regulate the details of morphogenesis in subtle ways, which have observable consequences on form.

The effects of diet and the malleability of bone quality and geometry is most sensibly investigated in an intraspecific context (Bouvier and Hylander, 1981; Ciochon et al., 1997; J. Scott et al., 2014a, 2014b), although intraspecific responsiveness sometimes manages to transcend the scope of interspecific variation (Ravosa et al., 2016). A few uncontroversial conclusions can be drawn from the larger body of work on skeletal ontogeny: (1) a primate species' skeletal bauplan is generated *in utero*, meaning that (2) skeletal form is only partially shaped by species ecology experienced ontogenetically, from which it follows that (3) this form is buffered to some degree with respect to the extraorganismal environment. What this suggests for paleoanthropological framing of adaptive arguments is that, given their morphology, *Australopithecus* and *Paranthropus* sought out—chose among—alternative environments, and they made food choices and adopted foraging strategies that were well-enough served by their performance capacities.

Chapter Six

The Osteocyte Perspective
on Human Evolution

Paleontological functional inference can benefit from the underutilized perspective of bone tissue and its developmental lability. Although reductionist in scope, this adds depth to a functional analysis, in that tissue-level responses to environmental stimuli are explicitly considered. Current orthodoxy is that these responses are finite, if not fully predictable, and different results among experimental studies implicate variables other than strains as important in determining the form of the adult skeleton. The utility of these responses depends on the ability of the osteocyte network to perceive the stress environment with some fidelity. The nature of this cellular perception remains to be fully elucidated (Vashishth et al., 2000; Riddle and Donahue, 2009; Pivonka et al., 2018), but I will provide examples to illustrate how an osteocyte's "point of view" can inform hypotheses of fossil function and evolution. Finally, I will offer recommendations on how this developmental perspective can inform a fossil record in which ontogenetic data are sparse.

The inference of adaptation, from the perspective of bone biology, as opposed to ecology, presents its own set of problems. Chief among these is that bone metabolic activity (modeling and remodeling) represents a domain of physiological adaptation,[1] as opposed to historical adaptation in the sense of Gould (1977) and Lauder (1981). The idea is that there is permanence to historical adaptation in terms of species-specific features that are inherited, with the corollary that phenotypic plasticity will not override the expression of the adaptation if that trait still functions in the manner under which it was selected. The view of Strait et al. (2013) is that physiological adaptation can be ignored if hypotheses of evolutionary adaptation make identical mechanical predictions. This would suggest that their "ontogenetic mechanisms of skeletal adaptation" (Strait et al. 2013:344) ought to be pursued in paleoanthropology.

There is no argument that the capacity of organisms to model and remodel their skeletons over the lifespan is heritable. This being the case, the distinction between physiological and evolutionary (= historical) adaptation is, to some degree, artificial. The disagreement among paleoanthropologists revolves around the question of whether physiological adaptation has any major impact on evolutionary outcomes. The existence of recognizable (if fuzzy) species boundaries, in terms of morphology, indicates that the scope of physiological adaptation over the lifespan is finite. It is also helpful to recognize that physiological adaptation is a subset of the broader phenomenon of phenotypic plasticity. Many physiological adjustments over the lifespan are helpful, but there are plenty that are not. The remodeling process can be seen as a plastic response to temporally proximate conditions, but osteoporosis illustrates a cost of plasticity. For plasticity to be adaptive in the broad sense, the appropriate environmental cues need to be sensed by the organism, so that the physiological responses are appropriate for present conditions. For example, senescence of signaling capacity in the osteocyte network is arguably more consequential than inefficiencies of osteoblast recruitment and performance.

The worry that a focus on morphogenetic variation is an insidious foray into a Lamarckian paradigm—i.e., by suggesting that there is an evolutionary role for adjustments, often ephemeral, over the lifespan—is misplaced. In the 19th century, August Weismann chopped off the tails of mice for several generations to disprove the idea that the induced "tailless" trait expression would be inherited. While this may have provided the most unambiguous outcome in the history of experimental biology, the study design was inadequate for its purpose. It was akin to arguing that falling off a ladder is an instance of plasticity.

There have been experimental manipulations that have attempted to illuminate the mechanobiology of locomotor morphology. The forced adoption of bipedality in species that ordinarily do not practice it results in modifications of the skeleton that indicate a facultative adjustment to novel stresses and posture (Nakatsukasa et al., 1995; Russo et al., 2018), but in no case does this skeletal modification converge on the morphological details of the human skeleton. Developmental plasticity does not appear to be generally capable of producing the scale of morphological changes that one would otherwise regard as recognizable species- or clade-level adaptations to distinct behaviors. On the face of it, this would imbue a greater role for historical adaptation in driving morphological evolution.

The perceived impotence of physiological adaptation is misleading. Scenarios of the origins of bipedality suggest that physiological adaptation is not only complementary to, but necessary for the production of historical adaptations. Marks (1989) noted that the conventional scenario for acquisition of bipedality involved the faculty having preceded the genetic underpinnings of the obligate behavior. Once this assumption has been made plain, he argued, the Lamarckian undercurrent prompted a change in emphasis to a "genes first" model. This presents two possibilities—both unpalatable. The first is the "hopeful monster," in which large-scale, essentially instantaneous changes to the genome produce the necessary morphological modifications for competent bipedalism. The second is the selection of alleles that incrementally fix traits favorable for bipedality. In Marks's (1989) view, this model was an appeal to a gradualist sensibility, forcing an assumption that the piecemeal adoption of locomotor behavior could proceed without difficulty. In any case, both hopeful monster and incremental selection models posit that the evolution of bipedality requires the requisite genes being there prior to the behavior. This entails a few difficulties, including the charge that the genome is imbued with precognition. This need not be a serious concern, however. The explanation behind canalization (and the heritability of "acquired" morphology) is that there is latent genetic variation that is susceptible to selection. Cheverud's (1996:45) position that "developmental integration structures genetic integration" is particularly germane in this context.

Marks (1989) argued that the most plausible scenario was one of genetic assimilation: stem hominin adaptability was followed by canalization. Phenotypic plasticity was the catalyst. According to Marks (1989:497), "The developing organism is a reactive system upon which the habit of bipedalism can act as a stimulus. If early hominids possessed the ability to make such developmental modifications on adopting the bipedal habit, and if there were genetic variation for this ability, both in terms of the magnitude of the response and in the facility with which the response could be elicited, then selection could operate upon this cryptic genetic variation." His insistence that historical adaptation be conceptually distinct from physiological adaptation may be acceptable from the standpoint of effective communication, but this does not mean that the concepts are processually independent.

There can be no doubt that fossil morphology is the product of some combination of physiological and historical adaptation. It also bears repeating that if the goal of a research program is the identification of traits that were produced by natural selection, then a focus on historical adaptation is appropriate. If the goal is reconstruction of behavior, however, one should also be interested in considering biological roles, not all of which will represent adaptations in the conventional sense. An unresolved question is whether these roles are more accessible via a bone development perspective than by a matching of ecology and morphology. This is something of a false dichotomy, once ecology is thought of as being more nuanced than simply a fixed set of problems to be solved.

Bone Adaptation: The View from the Mailroom

Skeletal morphology, like anything else in biology, can be reduced to chemistry and physics, with idiosyncrasies of history thrown in. In small doses, this perspective is useful for functional morphology, since it can serve to identify what is physiologically possible (Alexander, 1985). The morphometric discrimination of primate species apparently indicates that the range of skeletal variation bumps up against some sort of limit that approximately coincides with species boundaries; at least this is the unstated hope of comparative methods. Thus, presumably, the genetic disjunction that defines species is largely responsible for the accompanying osteological markers that are invoked for taxonomic purposes.

The potential pitfalls of uncritically accepting this principle have been clearly recognized in paleoanthropology (Tattersall, 1986; Wood, 1992; Kimbel and Martin, 2013). Morphometric distance does not provide unambiguous insight into species designations. Even if one can justify a particular benchmark of distance in a given comparison, no one believes that this standard can be applied successfully across the board. *Cercopithecus* guenons generally are not easily discriminated by skull morphometrics, even though (as one would expect) correlations exist between genetic and morphological distances among them (Cardini and Elton, 2008). *Mandrillus*, *Theropithecus*, and *Papio* can be easily distinguished by osteological morphometric criteria, but a cogent argument can be made that they are best classified as subspecies, rather than as distinct genera, once covariances of geography and genetics are considered (S. Frost et al., 2003). By virtue of a "total morphological pattern," mangabeys fit the criterion of a monophyletic group, but they are more plausibly paraphyletic (Disotell, 1994; Fleagle and McGraw, 2002). A neat correspondence of morphological variation and taxonomy was more credible before genetic or clinal morphometric data was scrutinized (Cardini and Elton, 2009).

Despite these examples, species-specific skeletal morphology is generally identifiable among primates, either in terms of trait-states or statistically. Ascribing this fact exclusively to behavioral variation would be facile, in light of the outcome of a simple thought experiment. Members of different species could be subjected to identical loading regimens during growth, with the expectation that their skeletons would converge on the same morphology. I am confident that few anthropologists would bet on that result. On the other hand, the guiding principles of the Mechanostat or Wolff's Law are premised on (and are useless without) the idea that bone tissue behaves in predictable ways to external stimuli, *no matter who you are.* This unadulterated statement of the principle that form follows function conflicts with the observation of finite skeletal modification during development, as well as with the nature of variance in outcomes of experimental manipulations. The general explanation for this conflict is that the capacity to model and remodel bone, and the physiological dictates of bone growth,[2] are variable and heritable. That being the case, one expects those capacities to be more homogeneous (but certainly not invariant) within rather than between species.

It is important to delineate what, exactly, "capacity" entails in the context of bone development. The differing responses of bone to naturalistic and experimental treatments are what keep the skeletal biology journals going. For there to be a relationship between organismal activity and bone physiology, mechanotransduction is essential: bone cells must be able to somehow sense mechanical events if they are to initiate a metabolic response to them. Molecular signaling is the proximate means by which mechanical stimuli provoke a cellular response, but there are myriad other factors that can alter or override this response. A few examples (far from exhaustive) are provided below.

Hormonal influences are highly consequential. Post-menopausal osteoporosis is a major public health concern in human populations, and there is strong evidence that estrogen deficiency leads to increased osteoclast production (D'Amelio et al., 2008). In contrast, parathyroid hormone (PTH) enhances osteoblast activity, and pharmacological interventions (using a recombinant fragment of human PTH) show increased anabolism of bone, believed to be attributable to increased osteoblast recruitment and proliferation (Trichilo and Pivonka, 2018). These hormonal effects on bone activity may result from a resetting of the threshold sensitivity of bone cells to mechanical loads (Gunness and Hock, 1993; Li et al., 2003). It is easy—and misleading—to generalize from these observations, because the implication is that osteoporosis or other hormonally mediated conditions will manifest in systemic fashion. Ravosa and Kane (2017) note that osteoporosis is typically observed in the appendicular skeleton, rather than the skull. They suggest that this might be tied to local safety factors. Low safety factors should be associated with greater plasticity as a proximate failsafe, but at a cost of plastic responses that do not respond appropriately to certain environmental signals.

Nutritional effects on bone quality are well known; e.g., a dearth of dietary calcium has negative repercussions, which are exacerbated by vitamin D deficiency. The influence of diet quality and caloric sufficiency on bone metabolism depends, at a minimum, on the interactive effects of nutrients and hormonal activity. Rizzoli (2014) notes that dietary calcium is potentially involved in bone modeling, as well as providing a mineral source, and that protein intake is positively associated with variables reflecting anabolic activity. Despite the established influence of diet on bone tissue, isolating the effects of nutritional, endocrine, and physical activity levels on bone metabolic activity remains challenging from both theoretical and experimental perspectives.

A final example underscores the important point that bone behavior sometimes cannot be generalized across taxonomic boundaries. Rats, which serve as the model species in a wide array of biomedical research programs, are unsuitable for the study of secondary remodeling, because they do not ordinarily exhibit it (Lad, 2018). When rats experience a physiological (and behaviorally unrealistic) overload, however, a remodeling response can be induced (Bentolila et al., 1998). Obviously the signaling capacity is intact but suppressed or ignored in normal rat life history. It is also inarguable that the naturally occurring damage burden to rat bones can be tolerated without repair, so presumably the strategy of not recruiting bone cell lineages for remodeling purposes has no major structural consequences. Skedros (2012) cautiously offers that an allometric principle might also explain the absence of remodeling in small animals: there is a "stressed volume effect" that, in effect, immunizes small bones against fatigue failure. This benefit is based on what can be viewed as a statistical argument. Large bone volumes are more likely to contain consequential structural flaws, because there are more opportunities for design imperfections to arise. A systematic survey of diminutive mammals is badly needed to evaluate the force of this explanation.

The effects of endocrine, nutritional, and genetic variables are interactive and ubiquitous. Bone has been discovered to be instrumental in endocrine function itself (Fukumoto and Martin, 2009). From the standpoint of mechanobiology, the important message is that these factors will conspire to undermine a precisely tuned signal–response relationship for either bone modeling or remodeling. Missed predictions from a Wolff's Law or Mechanostat ideal should consequently be unsurprising.

The osteocyte's view of the environment is fundamentally different from that of the paleoanthropologist. Despite this being a vacuous observation, the distinction is crucial for formalizing a mechanobiological approach to paleontology and understanding both its potential and limitations. Depending on your perspective about the evolution of bipedality, climbing may or may not have had a critical role in determining fitness in early hominins. Fleagle et al. (1981) suggested that, biomechanically, climbing and bipedality share common recruitment patterns of the hip and thigh musculature. This implies that a major neuromuscular reorganization would have been largely unnecessary for acquisition of bipedality, and that, instead, skeletal modification was the major transformation. On the other hand, Lovejoy et al. (2009) proposed that the basal hominin was unlikely to have been a climber in the manner of hominoids today. Consequently, the transition to bipedality would probably have involved a more substantial musculoskeletal reorganization. This might mean a relatively rapid, punctuated transition. What does the perspective of the osteocyte have to say about these possibilities?

Testing Equifinality

If you are interested in identifying adaptations or reconstructing behavior from paleontological data, the stuff of your nightmares is equifinality. According to this principle, in open systems (which most certainly applies to evolutionary contexts), identical outcomes can be generated from distinct processes. For the morphologist, this means that a given trait or trait value can arise from disparate developmental histories or behaviors. As a result, otherwise easily distinguished environments could produce indistinguishable morphological patterns, or different algorithms of growth and development could result in statistically identical morphological variation. In both cases, functional inference is compromised.

In the realm of skeletal adaptation, one example would be the functional significance of axially symmetric, cross-sectional geometries of long bones. A number of hypothetical loading scenarios can be recruited to explain the functional utility of a perfectly cylindrical section in terms of either minimization or homogenization of stress. First, a simple load case of axial compression suffices, because any shape will do. Second, the optimum solution for a twisting load is a cylindrical configuration. Third, in the anthropological literature, cylindrical sections are usually interpreted as reflecting unpredictable and random bending regimes. The first possibility is dismissible, on the grounds that isolated axial loads are impossible to imagine during locomotion. The remaining two are a tossup. Some degree of torsion is probably present in most locomotor scenarios, and bending undoubtedly accounts for the highest normal strains. Nonetheless, the bending scenario must address two issues. First, if we can define species-specific locomotion as a finite set of stereotypical behaviors, the likelihood of a uniform distribution of multiaxial bending loads is logically quite low. Second, in selecting the bending model, we tacitly accept longitudinal strains as the decisive osteogenic signal. On circumstantial evidence alone, this is unlikely (Demes et al., 1998, 2001). If shear strains are the osteogenic stimulus, then both bending and torsion explanations make sense,[3] and the geometry of the bone section is readily explained by either (Skedros, 2012).

As a matter of practice, paleoanthropologists employ an essentialist functional perspective: hominin species are defined by unique constellations of features, some number of which reflect adaptations. Seldom is any one hypothesis acceptable to the paleoanthropological community. Cartmill (1990) warns that valid tests of these competing hypotheses may be hard to come by. Adopting a mechanobiological (or osteocyte's) perspective might lessen our pessimism. Yet how can a developmental perspective improve our discrimination in the fossil record? The following examples explore the ways in which mechanobiology can evaluate competing adaptive scenarios in human evolution.

Too Many Degrees of Freedom

One of the hazards of paleoanthropological inference is the oversized role that imagination can have on the formulation of hypotheses, as there is no direct preservation of behavior, habitat, or social structure that can constrain an argument. Dart's (1957) osteodontokeratic musings were eventually undermined by Brain's (1981) systematic investigation of South African caves, and taphonomy has since functioned as an important restraint on overinterpretation in human evolution. Does the mechanobiological perspective have a similar role to play for informing functional inference? An important test of this idea is its ability to resolve competing hypotheses.

Carrier and Morgan (2015) challenged the standard view that the peculiar facial morphology of australopiths is attributable to feeding specialization. They reasoned that because modern people break facial bones during bouts of interpersonal violence more often than other skeletal elements, hominins historically have been at risk for facial skeletal trauma, owing to our inherited penchant for fighting. Sexual selection explains the behavior, with males of dimorphic taxa being more prone to pugilistic encounters. Because no direct evidence of interpersonal violence was offered, their argument employed the questionable logic of "not A, therefore B." Specifically, they contended that dental microwear of fossil teeth contradicted one dietary hypothesis (hard-object feeding), and that isotopic evidence affirmed another (a grass- and sedge-based diet), both of which were inexplicable with reference to morphology.[4] In addition, if mastication or other ingestive behaviors were entirely responsible for facial mass and geometry, they reasoned, then the facial skeleton as a whole should experience a homogeneous strain field under such behaviors. This is a straw man, insofar as the heterogeneity of strain fields is the rule *in vivo*. There is nothing unusual about the australopith facial skeleton in this regard. A homogeneous "adapted" strain field exists only in the mind of the 20th century morphologist.

Since direct evidence for slugging conspecifics is hard to establish in the fossil record, one would need to have some way of estimating the frequency of physically violent encounters. This is obviated if the premise of the argument is that selection has produced faces that are largely impervious to fracture. More importantly, brutal encounters constitute one-off events that the osteocyte network can detect, but the absence of cyclical regularity suggests that mechanotransduction would have been relatively impotent with respect to promoting facial bone hypertrophy. Mastication would remain a far more effective and consistent osteogenic stimulus (regardless of diet), as one would assume that blows to the face would, on most days, provide a stress stimulus of zero.

One can retreat to the contention that physiological adaptation is irrelevant, invoking the rationale of Strait et al. (2013). This position maintains that safety factors are built in morphogenetically, independent of vagaries of the developmental mechanical environment. As suggested in Chapter 2, however, safety factors may not exist to prevent accidents. While they undoubtedly keep some fractures from occurring, if an agitated *Paranthropus* should employ a rock or a club effectively in an agonistic context, even the face of KNM-ER 406 (see Figure 5.1) would not withstand the insult. This argues against a scenario in which facial hypertrophy initially represents a plastic response and is subsequently canalized in australopith lineages. One must consider what osteogenic stimulus favors that hypertrophy in the first place, and it is not clear that repeated facial trauma will yield a more robust adult facial skeleton.

This need not be the stimulus if facial hypertrophy is socially, rather than mechanically, mediated (Fannin et al., 2021). Hormonal action is implicated in other examples of facial hypertrophy. For instance, male mandrills have osseous facial features that play no meaningful role in masticatory mechanics. The idea of signaling—an effective means to avoid violent conflict—should not be dismissed out of hand. The question is the cost of maintenance. Maxillary and frontal sinuses underlie massive primate faces, and these spaces compromise structural integrity, especially in the context of trauma. From the perspective of the osteocyte (as a regulator of skeletal mass and geometry), the australopith face is better seen as an apparent rather than an actual buttress.

This does not mean that facial hypertrophy could not evolve to mitigate effects of interpersonal violence, but the mechanobiological processes responsive to masticatory stress stimuli and facial trauma are not analogous, as far as bone cellular activity is concerned. The most plausible scenario would seem to be one of genetic assimilation, in which the most efficient structural strategy is simply to build a large face, irrespective of the frequency (or even presence) of traumatic stimuli.

The mechanobiological perspective cannot settle this issue, but it certainly informs alternative hypotheses. The major challenge to Carrier and Morgan's (2015) thesis is the role envisaged for postcanine megadontia. They saw large teeth as objects built to resist fracture under trauma, and as a relatively safe means to transfer energy from the mandible to the cranium. There is both a mechanical and a comparative reason to be skeptical of this argument. In the first place, the periodontal support of the maxillary molars in *Paranthropus* extends into the maxillary sinus (R. Leakey and Walker, 1988), completely undermining the stress-shielding role of the teeth. In addition, Carrier and Morgan's (2015) own survey suggested that molar fracture is a relatively uncommon corollary of interpersonal violence in modern people, in which dental reduction is a hallmark of our species.

Only People Speak, and Only People Have Chins

The chin is a unique attribute of modern humans and, as a result, has attracted a great deal of speculation as to its adaptive significance.[5] Some of the recurring explanations have included language, altered chewing mechanics, dental reduction, and sexual ornamentation. Pampush (2015) argued, based on evolutionary simulation, that a rate shift in symphyseal morphology—indicative of a selective event—occurred late in the evolution of *Homo*. Alternatively, Coquerelle et al. (2013) contended that chin morphology is compelled by spatial restriction during facial growth in *Homo*. Gould and Lewontin (1979) invoked the chin as a signature example of a nonadaptation, based on the logic that alveolar resorption during development was sufficient to explain the structure as a trait that was previously masked and was therefore never "transformed" by selection. These developmental explanations are not sufficient, in themselves, to identify the presence or absence of adaptation, because development suffices as a proximate rationale for trait expression, irrespective of utility. If this is a valid negation of adaptive significance, then an understanding of how genu valgus, hallucial adduction, and lumbar lordosis emerge ontogenetically is adequate to deny adaptive status for bipedality.

Unsurprisingly, adaptive explanations for the human chin proceed from identification of some second unique human feature that "covaries" with it. If the chin is seen as a discrete character, rather than a continuous one, then any number of adaptive scenarios can be proposed without much need or opportunity to offer a convincing test. Human spoken language is a communication system unlike any other known in nature, and, given the tongue's association with the mandible, a connection between language and the chin has been proposed several times, beginning in the early 20th century.[6] Du Brul and Sicher (1954), Demes et al. (1984), and Daegling (1993) viewed the chin as a load-bearing structure and therefore surmised that, from a functional-mechanical perspective, language could not have a tangible influence on it. Instead, the chin was seen as an outcome of the overall geometric change in the mandible late in human evolution. The loads experienced by the mandible remained the same, but the relative magnitudes and effects of the different loads were altered, and this change in the biomechanical milieu explained the chin (Figure 6.1). Weidenreich (1936) considered dental reduction in later human evolution to have accounted for the chin's emergence, reasoning that shrinking tooth roots permitted a reduction in alveolar support for the anterior dentition. Since the chin was simply the persistence of the basal symphyseal region, this explanation anticipated that of Gould and Lewontin (1979).

The first step toward resolution is to recognize that the chin's status as a "thing" that humans have and other animals lack is the result of an arbitrary decision about trait definition. If, instead, we take a less essentialist approach and see the chin as an extreme expression of a morphological continuum, then the functional-mechanical basis for its form can be considered in comparative perspective. This is what Du Brul and Sicher (1954) advocated when—perhaps out of exasperation—they settled on describing the chin as "a blob of bone"; i.e., explicable with reference to other such blobs found in the skeleton of our and other species.

The proposals of Du Brul and Sicher (1954), Demes et al. (1984), and Daegling (1993) were explicitly mechanically based. The idea common to them was that the identification of regions prone to tension would be subject to selection for either geometric alteration or cortical thickening, given bone's relative weakness under this type of stress. None of these possibilities undermined Weidenreich's (1936) idea that the chin is an artifact of labial alveolar bone recession. Demes et al. (1984) and Daegling (1993) suggested that the twisting of the mandibular corpora loaded

Adult Females

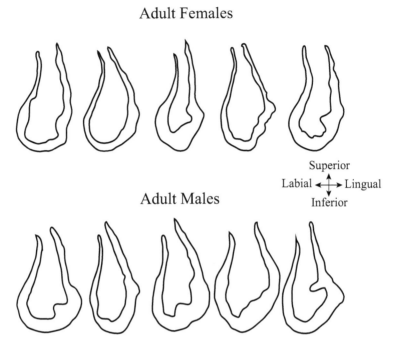

Superior

Labial ←↕→ Lingual

Adult Males

Inferior

Figure 6.1. Midsagittal cortical bone contours of the modern human chin. Arguments over the adaptive significance of the chin persist after more than a century of debate. Lack of resolution is due in part to issues of trait definition (see Schwartz and Tattersall, 2000). From a functional-mechanical perspective, the chin can be conceived of as a bizarre, but homologous, example of the mammalian mandibular symphysis. In this case, the labial basal protrusion—the mark of "chin-ness"—can be explained as an artifact of reduced alveolar support for a diminutive anterior dentition. The images in this figure also call for an explanation of the pronounced and uneven hypertrophy of cortical bone along the basal aspect. Explanations might be sought in mechanobiological terms; i.e., what intervals along the continuum of load frequency versus magnitude (see Figure 2.8) correspond to human activities? The idea that dimorphism in chin shape indicates an ornamentation function must explain why such a large metabolic investment is needed to support it; i.e., the chin would appear to be overbuilt if its biological role is signaling. From Pampush and Daegling (2016), used with permission from John Wiley and Sons.

the anterior mandible with a type of "coronal bending," in which the base of the symphysis was under tensile strain.[7] Du Brul and Sicher (1954) posited that the rearrangement of the overall proportions of the human mandible endowed the lateral pterygoid muscles with an enhanced capacity for producing a medially directed bending moment at the symphysis. This would result in tension along the chin's labial face during chewing, precisely where it protrudes. When their explanation was was proposed, it made good sense, but subsequent electromyographic study indicated that these muscles are not particularly active during chewing. They are more likely to be active during certain speech activities, however.

The idea that spoken language has a tangible functional-mechanical link to chin morphology has historically been met with skepticism. This dubiousness persisted because the idea than speech activity could produce anything larger than trivial stresses in bone was not credible, and there was always a more plausible alternative. Du Brul and Sicher (1954) and Coquerelle et al. (2013) foresaw a spatial dilemma looming, in that airway impingement or endangerment of cervical viscera would result from an ape-like developmental pattern (specifically, a lingually inclined symphysis with a simian shelf) in a highly flexed skull. Hershkovitz (1970) regarded this as an invented problem, since it postulated a developmental response to a spatial conflict that probably never arose in an actual animal. Another problem was the issue of how to model a load case for speech operating on the mandible, as no *in vivo* data to facilitate validation exist. Ichim et al. (2007) produced a pair of finite element models in which the form of the human chin reduced labial strain gradients relative to a "chinless" analog. The inference was drawn by considering the action of the genioglossus in isolation—a completely unrealistic load case. The genioglossus is highly active in mammals, generally to maintain an open airway, and it is not clear that its recruitment during human speech adds a significant stress stimulus. This is a case of high model specificity that is compromised by an inattention to realism.

The reservation that bone strains imparted by speech could not be osteogenic is not entirely justified. As Figure 2.8 suggests, small strains may be potent stimuli for bone modeling if applied at high frequency, and speech certainly fits that combination, compared with locomotor or masticatory loads. The strange property of the human chin is that, compared with homologous locations in nonhuman primates, we have an extraordinary amount of cortical bone in this region (Pampush and Daegling, 2016). This undermines myriad hypotheses, or at least encourages qualifications. Hershkovitz (1970) and Thayer and Dobson (2010) argued that the chin was selected as an ornament related to sexual selection. Cortical hypertrophy is an unnecessary and costly covariate of chin shape. Given the universality of extraoral food preparation and small masticatory muscle mass in modern humans, comparatively low strains in human mandibles are expected, *ceteris paribus*. This suggests that added cortical thickness is, from a perspective of stress management, a waste of metabolic resources. These utilitarian constructions fall short, but perhaps a general mechanobiological process explains what we see. Assuming that (1) speech produces a stress stimulus that occupies the low-load, high-frequency end of an osteogenic continuum, and (2) speech is practiced with sufficient frequency that speech events are sampled with fidelity by the osteocyte network, is it possible that the stress stimulus of speech produces comparatively high levels of osteogenic modeling activity?

While there is no compelling argument that the shape of the chin is a mechanical adaptation to speech, one aspect of chin morphology—cortical bone mass—is perhaps explicable by the mechanical effects of the high-frequency oscillations produced during speech. This reasoning is actually nonadaptive with respect to the activity of speaking, and the anabolic response is, by all appearances, unnecessary. What may be occurring is that the osteocyte network is responding to a stress stimulus, exactly as it has evolved to do, but the stimulus is novel, and, in this case, the skeletal response is out of proportion to the functional need (Daegling, 2012). No one has ever broken their jaw by speaking too much, even if we wish that this were so. A reexamination of Figure 3.7 should discourage wholesale acceptance of the idea that speech begat the chin. If, in fact, saturation of the mechanical signal is driven in the same fashion by load frequency under high or low physiological strains, then the likelihood that speaking could induce measurable osteogenesis seems unlikely, because the effective stress stimulus would be unremarkable.

In this example, the mechanobiological response can be visualized as a programmed one that does not distinguish among activities that may or may not be harmful down the road. In vertebrate history, this response, more often than not, is helpful. Inefficient, but not particularly deleterious, responses should be unsurprising. Under this scenario, our mandibles are full of bone that we don't need. This is an imperfection—a failure of evolutionary optimization—but it is obvious that the costs are not particularly grave. The intriguing question that presents itself concerns the generality of mechanobiological responses. How much are these conserved across clades, species, and populations? It is already certain that populations experiencing similar load histories can differ in their metabolic responses (I. Wallace et al., 2012). A remaining question is where in the chain of strain–signaling–response the differences lie.

As noted above, the foregoing discussion exposes an apparent paradox with respect to signal saturation, as well as to how a daily stress stimulus is transduced into an effective stress stimulus (see Figures 3.6 and 3.7). High-frequency, low-magnitude loads create strain rates that should have osteogenic effects. These kinds of loads may be omnipresent, even during periods of apparent inactivity (due to tonic muscular activity, minor postural adjustments, etc.). If, as is widely believed, osteogenic signals decay rapidly after the onset of activity, as a function of cycle number, then it stands to reason that almost all effective strains (i.e., those sampled by the osteocyte network) will involve rapid cycles and small strains. Indeed, the stress-stimulus simulations depicted in Figures 3.6 and 3.7 sample from a theoretical loading spectrum that assumes vigorous activity of the kinds measured during locomotion or mastication. As noted for those simulations, variance in the effective strain stimulus is largely determined by the nature of loads immediately following a period of rest. These rest bouts, however, do not represent times when strains (and strain rates) are entirely absent. What this suggests is that small strains are not interpreted identically to large ones, even if the strain rates involved are comparable. The refractory period following a loading bout has a large influence on osteogenesis, and, in many contexts, the osteocyte network is recovering its mechanotransduction capacity in the presence of lower-magnitude peak strains. There is much more to learn, and what is critical is understanding what kind of information osteocytes are paying attention to.

Theoretical Morphology of Bone Growth: Shear Strain as the Architect

A two-part unresolved question concerns what aspects of the stress and strain environment are responsible for osteogenesis, and whether these are uniform across or even within species. It is unclear whether certain components of the strain tensor represent more important signals to the osteocyte network. One can recruit experimental evidence to support the idea that tensile, compressive, or shear strains are important osteogenic stimuli. Activity influences bone growth in measurable ways, but bones develop into their characteristic form even in the absence of physical activity. The amount of variance in bone growth that is attributable to the mechanical environment is unknown; we only know that it is present. Genetic variation undoubtedly has measurable effects on mechanical sensitivity (Robling and Turner, 2002).

Theoretical morphological models of skeletal development have underscored the value of a process-based perspective of bone adaptation (Carter and Beaupré, 2007; R. B. Martin et al., 2015). Such models represent hypothesis-generating methods for understanding the mechanical underpinnings of skeletal variation within and across species. At their most basic level, such models delimit a range of the possible in terms of morphology. More importantly (from the perspective of the functional morphologist), these models can describe how distinct algorithms of bone growth, with and without dependence on nonmechanical factors, impact morphology. A simple application is outlined in Figure 6.2.

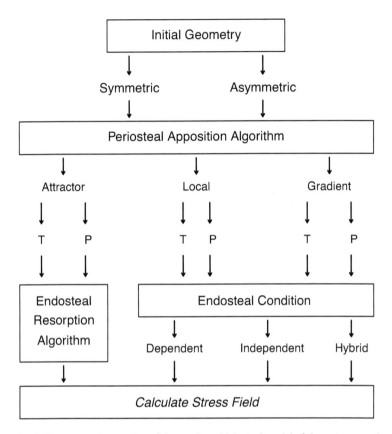

Figure 6.2. A diagrammatic overview of the mechanobiological model of shear-stress mediated bone growth illustrates the various model iterations. Attractor algorithms differ from the others in that the response to stress along the endosteal surface is not directly tied to periosteal activity or lack thereof: quiescence at one surface may not indicate quiescence at the other. The endosteal surface may be resorptive or quiescent, while the periosteal surface may be appositional or quiescent. This allows for a fine-tuning of the modeling response to maintain proximity to a target stress. The attractor, local, and gradient algorithms are implemented as binary and proportional responses to stress levels. The binary response involves a threshold stress (T), above which maximum apposition rates occur and below which there is no apposition at all. Under proportional responses (P), the apposition rates are tied to stress magnitudes relative to a target value (the threshold stress of the binary response); i.e., apposition rates are small in proximity to a target stress. In the gradient algorithms there is no target stress (see Table 6.1). Each of the local and gradient algorithms are evaluated against different sets of endosteal responses. A dependent response means that endosteal resorption is a fixed fraction of the rate of periosteal apposition. An independent response is a time-dependent growth function that is not reactive to stress levels but simulates somatic growth mediated by non-biomechanical developmental factors. This response involves a rapid decrease in weekly resorption rates until a time period representing sexual maturity. Finally, there is a hybrid condition in which the independent growth function is operative early, in concert with the stress-dependent function; the growth function declines to near zero by 100 weeks.

This model (Daegling et al., 2021) is loosely derivative of the approach published by Nowlan and Prendergast (2005), which simulated the evolution of bone cross-sectional geometry through varying ontogenetic programs. The present model is purely developmental, but it departs from Nowlan and Prendergast (2005) by utilizing a formula of stress calculation that is less restrictive geometrically; i.e., bone cross-sections are free to vary in shape.[8] Unlike most studies of primate bone mechanics, this model ignores normal bending strains and focuses exclusively on shear induced by a torsional load. The model assumes that load history is a simple function of body mass, which, in turn, is a time function (German et al., 1994). Load magnitude represents a shear stress stimulus.[9] One question the model asks is whether responsiveness to shear stress alone will produce adult bone morphology that is recognizable as such in terms of bone mass and geometry. The shear-resistance priority hypothesis (Skedros, 2012) posits that mitigation of shear stress represents the selective goal of bone metabolic activity. The rationale for the hypothesis rests on certain observations: (1) bone is relatively weak in shear, (2) shear stress is ubiquitous in many regions of the skeleton, and (3) the osteocyte network has the capacity to sense, and therefore respond to, shear strains.

The model I describe here was originally employed as a growth model for non–long bones; i.e., bones with small span:depth ratios, such as the calcaneus, mandible, or femoral neck.[10] Two initial elliptical geometries are employed: one involving symmetry of cortical geometry, and the other with marked asymmetry of the cortical shell (Figure 6.3). Growth is mediated by contrasting algorithms of an attractor state, local responsiveness, and responsiveness to stress gradients. The last of these ties bone apposition to the rate of change in stress as a function of distance along a bone surface, with higher rates inducing greater responses. Strain gradients are hypothesized as one possible trigger for bone apposition (see Hylander and Johnson, 1992; Gross et al., 1997; Judex et al., 1997). Most theoretical simulations model bone responsiveness as a local phenomenon; i.e., the magnitude of local strains dictates the presence and strength of a metabolic response in the same vicinity. Table 6.1 summarizes the specific attributes of these algorithms. The attractor state differs from the local response algorithm only in terms of the relationship between responses on the periosteal and endosteal surfaces. Each algorithm is modeled in terms of two types of responses: one involving a threshold response, and another in which the cellular response is proportional to the stress stimulus. These responses contrast the effects of discrete versus continuous functions in describing bone metabolic activity.

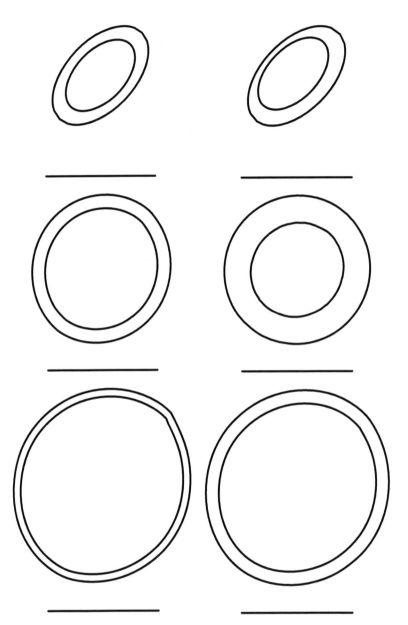

Figure 6.3. Initial geometries of the shear stress growth model (*top row*). These initial shapes have identical amounts of stress-resisting material but differ only in the offset of the internal contour. The maximum stress at week 1 was made equivalent between the two geometries. The adult geometry virtually "grown" from these initial shapes is shown for growth simulated under the the attractor stress algorithms (threshold, *middle row*; proportional, *bottom row*). The near-axial symmetry of the adult bones represent an approximation of an ideal engineering solution to the hypothetical load case that is imposed. Scale bars are provided for comparison.

Table 6.1   Algorithm attributes

A. Periosteal apposition

Stimulus	Response	Apposition rules
Attractor	threshold	$0 < 18\text{MPa}$, $0.5*\text{max} < 27\text{MPa}$, $\text{max} > 27\text{MPa}$
	proportional	$0 < 18\text{MPa}$, $(\tau/\tau_{max})*\text{max} < 27\text{MPa}$, $\text{max} > 27\text{MPa}$
Local	threshold	$0 < 18\text{MPa}$, $\text{max} > 18\text{MPa}$
	proportional	$\tau * 0.0004 < 54\text{MPa}$, $\text{max} > 54\text{MPa}$
Gradient	threshold	max when $g > g_{median}$, otherwise $(g/g_{median})*\text{max}$
	Proportional	$g/g_{max}*\text{max}$

B. Endosteal resorption

Stimulus	Response	Resorption rules
Attractor	threshold	$\text{max} < 18\text{MPa}$, $0 > 18\text{MPa}$
	proportional	$\text{max} < 18\text{MPa}$, $1 - (\tau/27\text{MPa})*\text{max} < 27\text{MPa}$, $0 > 27\text{MPa}$
Local	threshold	$0.333*\text{max}$
	proportional	$0.333*\text{max}$
Gradient	threshold	$0.333*\text{max}$
	Proportional	$0.333*\text{max}$

Notes: Among rules, "max" is the maximum apposition rate observed for lamellar bone (R. B. Martin et al., 2015); "τ" is the shear stress at a given location; and "g" is the strain gradient, defined as the change in stress over a unit distance on the periosteal surface. The value of 18 MPa represents a safety factor of 3 in shear, and 27 MPa indicates a safety factor of 2 (54 MPa is yield stress). Local and gradient stimuli were evaluated under three different endosteal conditions. The dependent condition is indicated in the table (*part B*); the independent condition is an exponential growth function in which resorption declines precipitously until sexual maturity (~ week 100 of growth) and then declines slowly thereafter. As the name suggests, this condition behaves independently of the stress environment. The hybrid condition is an interaction between the dependent (stress-sensitive) and independent conditions, in which the independent component is dominant early in growth and the dependent component becomes relatively more influential through time. A primary motivation for simulations under these different conditions is that the nature of the dependence of endosteal behavior (*part B*) to periosteal conditions (*part A*) is not well understood. Under the threshold gradient response, the gradient was calculated over 3 contiguous locations (points); under the proportional gradient response, the gradient was calculated over 5 contiguous points. Each perimeter consisted of 100 points.

The local environments of periosteal and endosteal regions *in vivo* are not similar. Endosteal surfaces have more direct communication with cells and tissues of the medullary cavity, while the periosteum adheres to the outer bone surfaces. Bone lining cells (quiescent osteoblasts) are present on both surfaces. During growth, there is apparent coupling of periosteal expansion (bone apposition) with endosteal expansion (bone resorption), but it is not obvious whether this is a coordinated response governed by mechanical stimuli. For this reason, endosteal modeling was evaluated under a variety of conditions. Under the attractor state algorithms, the endosteal modeling rate was allowed to respond to stress independently of the periosteal response. This is expected to yield a fine-tuned reaction to the stress stimulus, which conceivably will approach an optimal solution by one or more mechanical criteria (see below). For the local and gradient algorithms, the endosteal modeling was assumed to behave alternatively as (1) *dependent* on the periosteal response, (2) a growth function *independent* of stress and periosteal response, and (3) a *hybrid* with independent ("biological," or somatic, growth) and dependent (mechanically mediated) components. Under the independent condition, endosteal expansion follows an exponential growth function that declines rapidly with age (until skeletal maturity) and is insensitive to the stress environment. In the dependent condition, endosteal expansion tracks periosteal expansion at a rate that is one-third that of the periosteal surface. This condition approximates rates suggested from experimental manipulations (van der Meulen et al., 1993; Carpenter and Carter, 2008) and is consistent with observations that long bone cortices tend to thicken during growth. In the hybrid condition, a biologic growth function predominates early on, with the mechanical function assuming greater importance with age.

These growth algorithms are mechanically deterministic, such that, from alternative initial geometries, the adult form is more or less preordained (Figures 6.4 and 6.5), provided the modeling response is adequate to stave off structural failure. Model runs were terminated if ultimate stress was reached during a growth simulation. Loads proportional to body mass (using macaque growth data) were "applied" weekly until an age of 10 years.[11] These loads were allowed to vary randomly by up to 15% of the weekly load point estimate, to provide a source of developmental noise during model runs. Bone apposition rates were not allowed to exceed estimates of the maximum for lamellar bone formation (R. B. Martin et al., 2015).

Results of these growth simulations are summarized in Figures 6.6, 6.7, and 6.8, as well as in Table 6.2. There are different ways to evaluate the biomechanical utility of particular growth algorithms, and each of these can be viewed as "adaptive," based on a specific design criterion. Three criteria are evaluated: (1) a *stress-similarity* condition, in which the maximum stresses are unchanging throughout growth; (2) an *isostress* condition, in which the stress field is homogeneous; and (3) a *material economy* condition, in which ideal stress magnitudes (reckoned as safety factors) are maintained with a minimal metabolic investment, as determined by appositional and resorptive activity on periosteal and endosteal surfaces, respectively.

The stress similarity criterion assesses whether the calculated stress is stable over time, as well as whether it settles on a value that corresponds to an adequate safety factor (here defined as being between 2 and 3). These safety factors repre-

Figure 6.4. Adult bone sections resulting from the local stress growth algorithms, shown here for asymmetric initial geometries (threshold and proportional responses shown in the *left* and *right columns*, respectively). Local responses to stress differ from the attractor state algorithm in terms of the specificity of endosteal resorption relative to periosteal apposition: in these cases the bone is structurally sound throughout growth, but it does not approach the isostress condition to the degree seen under the attractor state. Under the independent endosteal condition (i.e., in which resorption is purely a function of growth, *middle row*), bones that are both efficient and structurally sound are produced. Under dependent and hybrid conditions (*top* and *bottom rows*, respectively), the threshold condition produces bones that have a very heterogeneous stress field, owing to pockets in which resorption is arrested, which is due to low local stresses. Also noteworthy is the departure of adult geometry from symmetry under the torsional load case. Such asymmetry is often interpreted as reflecting a structural response to the patterning of predictable bending loads. Scale bars are provided for comparison.

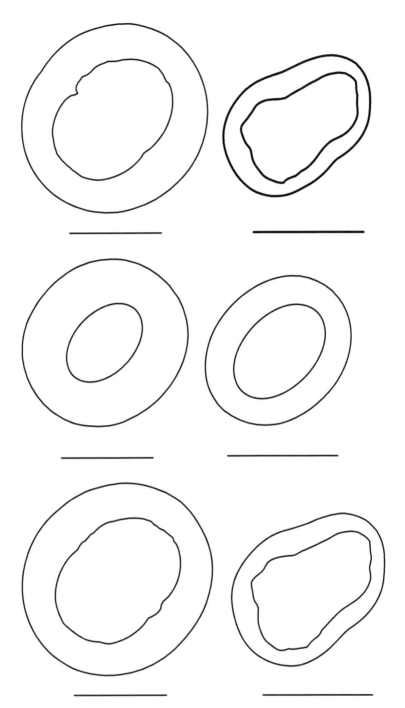

Figure 6.5. Adult bone sections resulting from the gradient algorithms, shown here for asymmetric initial geometries. Bones grown from the threshold algorithm are in the *left column*, those on the *right* are from the proportional algorithm. When strain gradients are the osteogenic signal, proportional responses produce odd geometries under dependent (*top of right column*) and hybrid (*bottom of right column*) endosteal resorption conditions, and safety factors are compromised during growth. Competent bones are produced under threshold responses, but they are relatively uneconomical, because they invest more bone than is needed to maintain structural integrity (see Table 6.2). Scale bars are provided for comparison.

Table 6.2 Model performance

Algorithm	Type	SF_{min}		SF_{final}		$Iso\tau_{min}$		$Iso\tau_{final}$		Apposition		Metabolic investment	
		sym	asym	sym	asym	sym	asym	sym	asym	sym	asym	sym	asym
Dependent endosteal condition													
Attractor	T	1.70	1.93	2.69	2.82	0.9	0.6	1.4	1.0	108.8	106.1	137.6	167.1
	P	1.98	2.14	2.73	2.69	0.5	0.7	1.1	1.7	197.3	188.7	335.2	343.5
Local	T	2.01	2.08	2.53	2.30	2.0	5.4	10.8	13.4	141.3	100.1	199.9	142.6
	P	1.33	1.52	2.12	2.31	2.9	16.2	16.0	20.3	121.0	91.5	158.0	120.7
Gradient	T	2.64	3.57	9.17	14.72	2.8	8.2	9.9	10.2	301.0	289.2	396.5	384.1
	P	0.94	1.22	1.89	1.85	5.5	12.9	13.7	22.1	118.4	91.5	164.8	127.2
Independent endosteal condition													
Attractor	T	1.70	1.93	2.69	2.82	0.9	0.6	1.4	1.0	108.8	106.1	137.6	167.1
	P	1.98	2.14	2.73	2.69	0.5	0.7	1.1	1.7	197.3	188.7	335.2	343.5
Local	T	1.77	1.54	2.75	2.80	2.7	1.4	7.3	5.7	115.6	83.1	137.8	105.3
	P	1.19	1.00	2.29	2.90	2.0	10.6	12.3	14.0	108.1	92.7	130.3	114.9
Gradient	T	2.16	2.27	7.75	13.67	4.2	5.7	6.1	6.0	220.4	219.3	242.6	241.5
	P	1.60	1.06	2.85	4.39	3.0	3.7	4.6	5.5	115.0	105.5	137.2	127.7

(*Continued*)

Table 6.2   (continued)

Algorithm	Type	SF$_{min}$		SF$_{final}$		Isoτ_{min}		Isoτ_{final}		Apposition		Metabolic investment	
		sym	asym	sym	asym	sym	asym	sym	asym	sym	asym	sym	asym
Hybrid endosteal condition													
Attractor	T	1.70	1.93	2.69	2.82	0.9	0.6	1.4	1.0	108.8	106.1	137.6	167.1
	P	1.98	2.14	2.73	2.69	0.5	0.7	1.1	1.7	197.3	188.7	335.2	343.5
Local	T	2.07	1.99	2.50	2.19	2.2	4.6	10.3	14.1	146.0	104.3	208.9	154.1
	P	1.37	1.51	2.18	2.27	3.2	16.1	16.3	20.8	127.4	97.8	172.4	135.2
Gradient	T	2.62	3.12	9.41	15.28	3.5	8.1	8.8	9.4	314.9	302.2	420.9	406.9
	P	1.57	1.51	2.76	2.27	4.1	11.1	14.0	20.2	153.7	109.4	218.1	149.8

Notes: Type "T" = threshold and "P" = proportional. The safety factor ("SF") is dimensionless ($\tau_{yield}/\tau_{observed}$); "Iso$\tau$" is an index of the isostress condition, which is the coefficient of variation of shear stress (tau, τ) for a given section (%); "Apposition" and "Metabolic Investment" are given in mm^2. "SF$_{min}$" and "Isoτ_{min}" are the lowest values for maximum stress and isostress condition observed over the growth period, and "SF$_{final}$" and "Isoτ_{final}" are the values at the conclusion of growth. Metabolic investment is the sum of the total apposition and total resorption over the growth period. Results are reported for initial symmetric ("sym") and asymmetric ("asym") geometries. Attractor algorithms are provided for comparison; their conditions for endosteal resorption are different from local and gradient algorithms. The expectation is that because the attractor algorithms have responses that are conditionally determined separately for endosteal and periosteal surfaces, their values will converge most closely on an ideal solution (safety factor of 3 and isostress of zero). Safety factors below 2 are suboptimal, because the bones are then susceptible to damage via physiological overload or fatigue. Safety factors above 4 can be regarded as inefficient, since there is more bone mass present than is required for structural integrity, meaning performance is necessarily compromised. Apposition and metabolic investment (apposition + resorption) are best evaluated relative to safety factors: the lowest value for each, given a suitable safety factor ($\sim$ 2–4), represents the most economical use of bone material for maintaining structural integrity.

sent a margin that would function to extend fatigue life and offer some protection against the odd accident. Stress similarity is also expected, based on empirical observations: most animals that have been examined exhibit peak skeletal strains during vigorous activity that correspond to these safety factors (Rubin and Lanyon, 1984a; Biewener, 1993). Stress similarity is never strictly observed in the model, but it is achieved for attractor algorithms and for the local threshold algorithms later in the growth period (Figure 6.6). The local proportional and gradient algorithms do not succeed in maintaining ideal safety factors, except for brief windows in the growth period.

The second measure of utility is the attainment of an isostress condition; i.e., one in which the stress magnitudes are the same everywhere in the section. This

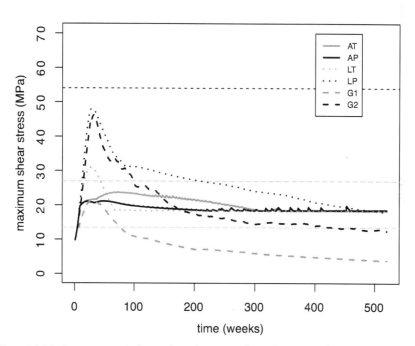

Figure 6.6. Maximum stresses in bones throughout growth are shown. The *dashed horizontal line* is yield stress; *gray dashed lines* represent safety factors of 2 and 4. AT = attractor threshold, AP = attractor proportional, LT = local threshold, LP = local proportional, GT = gradient threshold, and GP = gradient proportional. Ideally, bone moves toward and then maintains stress magnitudes consistent with ideal safety factors. The best means to achieve this is via threshold responses, using the attractor or local algorithms; proportional responses and/or stress gradients appear incapable of maintaining stress similarity throughout growth. The nature of endosteal dependence on stress signals has a relatively minor impact on the behavior of the different algorithms (see Table 6.2). The independent endosteal condition is depicted here.

represents a highly efficient deployment of bone tissue, because it means that a target stress is universally maintained. From the standpoint of optimization, any situation in which there is variance in stress indicates a waste of bone tissue. The isostress condition is, from an engineering standpoint, a desirable outcome, but there is little empirical evidence that this is what bone metabolic activity is aspiring to.[12] Still, from a thermodynamic or metabolic perspective, it represents a sound adaptive strategy.

Isostress conditions never develop under the various algorithms (Figure 6.7), but an attractor state algorithm, in which the endosteal and periosteal surfaces can

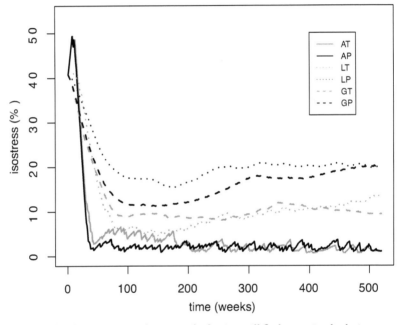

Figure 6.7. Under the assumption that natural selection will find an optimal solution to a biomechanical problem, growing bones should converge on an isostress condition; i.e., where stress magnitudes are identical throughout a structure. In the graph, AT = attractor threshold, AP = attractor proportional, LT = local threshold, LP = local proportional, GT = gradient threshold, and GP = gradient proportional. Under the load case specified in these simulations, this is theoretically achievable in the present context. The attractor algorithms largely succeed in reaching an isostress condition, because the endosteal resorption and periosteal apposition responses are independent and stress sensitive. All other algorithms, which are more biologically realistic, do not exhibit this capacity. The local and gradient algorithms are here plotted under the hybrid condition (i.e., a combination of somatic growth independent of stress and stress-reactive growth). If strain gradients are a primary osteogenic stimulus, an isostress condition would result in a complete absence of physiological adaptation.

respond independently, appears to be the best strategy for getting there. The near perfection of the final geometry, however, suggests that this is not terribly realistic (see Figure 6.3). Local algorithms are also fairly effective in achieving an isostress condition. Gradient algorithms are more volatile during growth and, under dependent endosteal tracking of periosteal modeling, actually may move away from an isostress condition.

The third design criterion under scrutiny is that of economical metabolic investment (Figure 6.8). This is measured simply as the sum of bone resorbed and bone added over the growth period, reflecting the energetic cost of recruiting and deploying populations of osteoclasts and osteoblasts. Here especially, the gradient algorithms perform miserably, investing in high levels of bone metabolic activity without functional benefit (meaning higher-than-needed safety factors). The local algorithms tend to add and remove bone most economically.

Obviously, under conditions in which endosteal expansion is a constant fraction of periosteal apposition (the dependent condition), stress at adulthood is relatively low, although both attractor state and local algorithms arrive at a value that is consistent with an expected safety factor. It is noteworthy that the partial or total decoupling of endosteal expansion from stress levels (the hybrid and independent endosteal conditions) produces bone geometry very similar to the dependent condition. This is not a demonstration of equifinality per se, but it suggests that there are very distinct means through which to grow a fully competent bone. The engrossing implication is that mechanical sensitivity of endosteal surfaces is not required for developing a functional bone. Gradient algorithms are only competent insofar as they do not break during growth. When stress gradients are responsible for the osteogenic signal, safety factors vary widely with time, ending up with needlessly large margins of safety but, more critically, also spending periods of time in which these safety factors are dangerously low.

The preceding observations underscore a largely ignored component of bone development: how efficiently is bone deployed for managing stress? One striking finding is that under both attractor and local algorithms, a threshold response is much more efficient at producing a competent skeleton, because it deploys far less bone in the aggregate, while maintaining a desirable safety factor. This indicates that a binary response to a stress stimulus—i.e., either maximum apposition or total quiescence—is a preferred means for managing the stress environment during ontogeny. This is consistent with the idea that bone cell recruitment operates in a quantum fashion (Forwood and Turner, 1994; Parfitt, 1994).

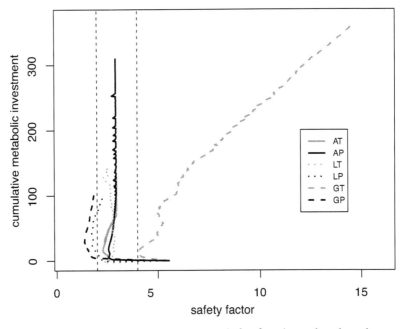

Figure 6.8. The maintenance of structural integrity (safety factor) as a physiological cost function (metabolic investment) illustrates the relative inefficiency of strain gradients as an osteogenic signal. The *dashed vertical lines* represent ideal safety factors between 2 and 4. Once again, AT = attractor threshold, AP = attractor proportional, LT = local threshold, LP = local proportional, GT = gradient threshold, and GP = gradient proportional. Ideally, bones maintain safety factors between 2 and 4 throughout growth. Local and attractor algorithms are capable in this regard, but strain gradients cannot maintain ideal safety factors over the course of growth. This is undesirable from both the standpoint of economy (high safety factors when responses are proportional) and that of integrity (low safety factors under threshold responses). Strain gradients have been correlated to the modeling response in the experimental arena, but the detection of gradients and the ensuing response *in vivo* may be quite different from the theoretical assumptions about those processes applied in these models. Even so, gradients—being changes in stress rather than stress magnitudes— may be a very indirect and inefficient means of maintaining adequate safety factors. The data depicted in this graph were generated from the dependent endosteal condition.

Given that this shear model ignores several biological factors involved in bone growth and employs boundary conditions of uncertain validity, what else can be learned from a purely theoretical mechanobiological exercise? First, this indicates that credible geometries and functional bones can develop if bone is exclusively responsive to shear (as opposed to normal) strains. But what is also compelling is that even in this situation, with a simplified load case and the absence of nuisance variables,[13] the algorithms do not produce bones that are ideal from an engineering perspective (with the sole exception of the threshold attractor state). Bending stress might be the mechanical underpinning of osteogenic responses, as is commonly assumed, but these models suggest that osteocyte sensitivity to normal (i.e., induced by bending) strains might merely be sufficient but not necessary.

Another insight—or, perhaps, more properly termed a perplexity—is that despite the experimental evidence that strain gradients represent an osteogenic signal (Judex et al., 1997; Hylander and Johnson, 2002), these signals will change rapidly and irregularly over short developmental time spans, such that they are highly ineffective, even in a fully deterministic model (Figure 6.8). This is somewhat surprising, since strain gradient severity correlates with sites of bone apposition (Gross et al., 1997). The different combinations of apposition algorithms and resorption conditions also provide a test case for equifinality, in terms of bone development. Despite conditions of identical initial geometry, load history, and load type, these algorithms produce very different adult morphologies, yet all of them follow quite simple rules that are directed toward maintaining structural competence.

It bears reemphasizing that this developmental model, as well as others assigning load history a deterministic role in bone morphology (Pauwels, 1980; Cowin et al., 1985; Carter et al. 1987), do not directly inform any adaptive hypothesis. What they do provide is an accounting of what is feasible and functional, given a finite number of assumptions. The advantage of such models is that they are quite malleable. For example, there are no serious impediments to introducing hormonal influences at specified times and adjusting bone turnover rates accordingly. These models can also be iterated to examine the effects of load case variability or a frequency-based stress stimulus on development. The important consideration is that the initial and boundary conditions be (1) explicit, and, to the extent possible, (2) empirically informed.

What can a simple developmental model such as this tell us about skeletal evolution? Focusing on the relatively poor performance of stress gradients—if, in fact, the osteocyte network utilizes gradients as the basis for osteogenic response—then there logically are a few competing explanations. One is that bone metabolic activity is not efficient in terms of maintaining stress similarity, homogenizing the stress field, or minimizing stress efficiently—even if growth factors independent of the stress environment are ignored. It is useful to ask how this state of affairs might have evolved, but pointless to insist that it could not have. (It is worth remembering that there are a multitude of ways in which a gradient can be transduced, and only two have been presented here.) In another possibility, by juxtaposing the gradient results against the attractor state, if natural selection should act to minimize stress efficiently, then there is some biological mechanism or historical condition that precludes an attractor state developmental program from being implemented. Theoretical models, by themselves, cannot answer these issues, but they do direct us toward novel research designs that will.

What Do Osteocytes Think About?

This question isn't to be taken literally, but it is a serious and important one allegorically. There is communication among osteocytes, but the content of the messages is fairly restricted by our own linguistic standards. We do not know what is being communicated or which aspects of the products of that communication have been the target of selection. Understanding this may solve fundamental questions of skeletal functional morphology. This will only come to pass by using experimental data to inform and adjust theoretical computational models.

If we assign a prominent role to the osteocyte as the sensory apparatus of the skeleton, it follows that the environment is perceived in rather simpler terms than those experienced by the sentient organism. For early hominins, behavioral decisions were based on processing special sensory inputs, and there can be no doubts that their cognitive capacity was proficient in interpreting the myriad differences between, for example, mesic (moist) versus xeric (dry) landscapes. But there is no reason to assume that this ecological information translates neatly down to the neighborhood of the osteocyte. If the frequency and intensity of activities changes with a shift in utilized landscapes, then perhaps a direct correspondence to the perceived environment could be achieved.

This can be illustrated by the following scenario. A hominin population moves into a xeric landscape from a more productive mesic one. There are several strategies for dealing with this novel environment, and two offer a relatively clear link from behavior to activity. The first is to seek out foods resources that were widely available previously but are now patchily distributed. This requires no change in feeding performance or investment, but it does necessitate an increase in traveling time as part of a daily activity budget. A second strategy is to opt for more spatially homogeneous but lower-quality forage. This involves daily travel equivalent to that with foraging in the mesic landscape, but feeding becomes challenging, with more time and effort each day in terms of ingestion and mastication. It is reasonable to assume that the osteocyte universe is capable of detecting these different environments, both in terms of the previously occupied landscape and in the alternative behavioral adjustments I have described. The environmental shift is manifested either as an increase in the stress stimulus on the locomotor skeleton (strategy one) or on the facial skeleton (strategy two).

Provided this shift is not too transient, a skeletal response to this changed environment is possible both developmentally (Ravosa et al., 2007; J. Scott et al., 2014a, 2014b; Menegaz and Ravosa, 2017) and over longer evolutionary time frames (West-Eberhard, 2005). The practical question is whether paleoanthropologists can detect this response by examining fossils, a possibility that has received far too little attention. Primate field studies are absolutely pivotal in this regard. What behavioral changes do populations undergo when moving among microhabitats? If there are behavioral shifts, does this necessarily lead to differences in functional activity, quantified as biomechanical variables? Another way of putting this is to ask, To what degree do the energetics of performance figure into behavioral decisions? Such approaches are both feasible and hypothesis generating (Careau and Garland, 2012; Sears and Angilletta 2015).

We remain fairly distant from understanding the limits of adaptive or behavioral discrimination in the paleontological context, but some obstacles are already known from the field of skeletal biology. As mentioned in Chapter 2, adult skeletal morphology is not a snapshot of contemporary physical activity. It probably reflects prior activity in the lifespan (Menegaz and Ravosa, 2017). This earlier behavior, ontogenetically speaking, is every bit as sensitive to natural selection as that of adults, but it is not necessarily the same in its details (see Doran, 1992). The amount of remodeling over the lifespan is potentially recoverable from fossil material, and the examination of secondary bone density in that material might provide insight into damage burdens in specific regions of skeletal elements. Complicating these inferences is the fact that secondary bone is not always found in regions where damage burdens are ostensibly high (Bouvier and Hylander, 1996; Lad et al., 2016). Moreover, there are no clear means to distinguish remodeling activity that was initiated for mechanical or metabolic purposes, beyond demonstrating an association with microcracks (Mori and Burr, 1993; R. B. Martin, 2002; Parfitt, 2002).

This paints a pessimistic picture of the scope of functional inferences in the fossil record. Yet a bleak view is not necessarily unhelpful. There is no need to abandon top-down ecological inferences in an effort to understand human evolution. Returning to Figure 3.2, models operating at different levels of analysis, from tissue to lineages, utilize distinct variables that, to some degree, cannot be compared across scales. On the other hand, tissue-level models that are congruent with bone biology (as presently understood) should, at the very least, *not* contradict higher level models that are using the same bones (and fossils) to characterize paleobiology. In other words, there ought to be consilience across levels, even if the precise questions being asked are not the same. This is not paradigmatically revolutionary in any way; paleoanthropologists do this informally all the time in defending a position. The only novelty is adding the caveat that models should be consistent with how bones actually respond to environmental stimuli. This is quite different from advocating for parsimony, as has traditionally been done in this field (Langdon, 1997). Consilience seeks an absence of contradiction, whereas parsimony seeks simplicity—at times despite contradiction.

Consilience sets a high bar, and it also builds in the linchpin of what some consider an anachronism: the criterion of falsifiability. Testability absolutely constrains the scope of hypotheses concerned with human evolution, but it is legitimate to ask, Of what use is a model that is resistant to meaningful refutation? Jablonski and Chaplin (1993) hypothesized that chest-pounding and other display behaviors precipitated habitual bipedality. Such activity is invisible paleontologically. Thus the model is unassailable, because no contrary evidence can be produced.[14] Its only defense is that all competing models, which offer predictions, can be shown to contain flaws. This is the "creation wins, evolution loses!" strategy of persuasion (Ruse, 1982). The undesirable property of the model is that it leaves nothing to build on. No new questions about human evolution emerge from it. Even if true, this is not useful. Falsification remains essential to our enterprise.

Expanding the Prescription

What functional or evolutionary morphology should aspire to in paleoanthropology is linking morphological innovation and variation to fitness by means other than a narrative. Adherence to the Kay and Cartmill (1977) prescription (see Chapter 3) is limited insurance against misguided speculation, but the cost is that there are aspects of paleobiology that are out of bounds analytically. That cost is unbearable if functional hypotheses are completely untestable and incapable of circumscription by ancillary data.

The dilemma can be analogized to the statistical concepts of Type I versus Type II errors. Type I errors are false positives: seeing significance when there is nothing of interest going on, other than sampling error. In paleoanthropology, as in statistics, one never really knows when a Type I error is in play. The Kay and Cartmill (1977) prescription is conservative, because its application will minimize Type I errors, and we are less gullible as a result. Type II errors occur when there really is something intriguing (i.e., "significant") going on, but we fail to recognize this. In this situation, we mistakenly accept a null hypothesis. Truzzi (2007) made the unconventional, if not heretical, suggestion that minimizing a Type II error was paramount if missing that provocative signal was highly consequential. It is likely that many practicing paleoanthropologists would describe their investigations in such terms.

The complement to a Type II error is the concept of statistical power, which depends on sample size and variability in the data. If power is high (say 95%), then Type II error is low (5%). By convention, the maximum tolerable Type II error is 20%. This means that we value low Type I errors over low Type II errors. Nonetheless, there is a way to minimize Type II errors: collect more (large N) and better (low variance) data. It is questionable that biological—and especially paleontological—data will cooperate in such an endeavor. Perhaps adjustments in methodological perspectives are more practical.

Bastir (2018) advocates a holistic perspective, albeit one that is not in the nebulous form that Williams (1992) disparaged. In Bastir's (2018) prescription, there are clear echoes of Dullemeijer's (1980, 1985) multiscalar perspective, as well as Moss's (1968, 1973, 1997a, 1997b, 1997c, 1997d) more atomistic sensibilities of the functional matrices. A system (e.g., the skull) is a valid object of investigation, but it is incompletely understood unless the infrasystem (Moss's "skeletal units") is scrutinized as well, and the influence of this infrasystem on the super-system (the organism) is also examined. This is sensible in the abstract, but the challenge will be to operationalize this holism across contexts. Bastir (2018) conceptualizes the approach from the standpoint of morphometric data, but Bookstein (2018) avers that these data do not translate neatly into an evolutionary biology framework. An explicit effort to situate morphometric data into formal models of modularity and development (e.g., Atchley and Hall, 1991) can allay these concerns. But for a vibrant evolutionary morphology to be realized, ethological, ecological, and environmental data must be recruited as well.

Rose and Lauder (1996), in eulogizing the demise of pan-adaptationism brought about by Gould and Lewontin's (1979) assault, also argue that recruitment of diverse kinds of data can effectively solve many of the ills of adaptive inference that troubled Williams (1966). In effect, their recommendations do not focus on definitions, terminology, and theoretical minutiae, but instead detail how distinct methodological approaches that document behavior, development, and phylogenetic transformations offer hope for a mature science of morphology. Building on some of these ideas, I suggest certain reorientings of research foci along the following lines.

Revisit load history from the osteocyte perspective. My critique of paleoanthropology is focused on the belief that bone bears the marks of activity, and behavior can be discerned, even if dimly, from those marks. Skepticism about the validity of this concept has to be balanced against a mechanobiological fact: bone is a tissue that is reactive to its physical environment in nonrandom ways. The principal problem has been that "activity" is conceived of in terms of a taxonomy created by paleoanthropologists, without regard for what that activity looks like to an osteocyte.

Skedros (2012) offers a multiscalar, multifactorial approach to reconstructing load history that includes cellular and tissue vantage points. It embraces several interacting hypothetical positions, two of which—the stressed volume effect and the shear-resistance priority hypothesis—have been introduced above. In addition, hypotheses relating to secondary osteon morphology (size, geometry, orientation, collagen fiber organization), load predictability, and complexity, as well as a "division of labor" explanation for modeling and remodeling activity, among others, are all brought to bear on the question of how bone tissue reacts to the immediate mechanical environment. This is the point of view that matters for understanding the evolution of the skeleton.

The significance of Skedros's (2012) proposal is that the bar is set quite high for validation, because there are multiple sources of data that must be consilient for details of load history to be seen as credible. There are, as he notes, a not-small number of studies that challenge several of the theoretical premises of the approach as a whole. The specificity of predictions, however, ensures that each aspect of this proposal is testable. I suspect at least one reason why much of the data is apparently contradictory is that the model species used in experimental research—i.e., the sources of potentially supportive data—have been quite diverse, including people and nonhuman primates, ungulates, rabbits, hyraxes, rodents, and carnivores. These reflect taxonomic differences at the ordinal level. Moreover, although we might wish to have a singular set of processes characterizing bone physiology in mammals, histologically speaking, it is already abundantly clear that variation in bone morphogenesis across species is not exceptional.

Examine the variance as well as the mean. Van Valen (1965) demonstrated that high morphological variance within species should prompt consideration that a taxon is eurytopic. This productively challenges the paleoanthropologist's essentialist tendencies. Potts (1998), in arguing for variability selection as the overarching process of human evolution, recognized this, but he did so from a behavioral perspective that did not alight specifically on morphology. Large variance in a trait might not mean weak or absent selection, but it could indicate multiple selective agents acting on different subsets of the population.

Williams (1992), Wagner and Altenberg (1996), and Hallgrímsson and Hall (2011) all pointed out that uncovering the mechanisms for generating variability effectively resolves the paradox of the ubiquity of natural selection and the universality of variation at all taxonomic levels. With respect to paleoanthropology, no doubt the inertia of essentialism has fostered resistance to the idea that attention to patterns of variance is helpful to the practice of functional morphology.

Articulate additional biological roles and their costs. This is particularly germane for primates, hominins in particular, as their cognitive capabilities expand the universe of potential biological roles for morphological traits. If the role entails a functional accommodation, then the morphology governing function for the primary (presumably selected) behavior is not at an adaptive peak in terms of performance. The compromise entails a performance cost to the primary function of the focal trait, but the associated biological role(s) is or are presumably sufficiently critical that the cost is bearable, or even indispensable. These possibilities can only be indirectly tested via neontological observation, but the cost function can be evaluated through a variety of means, including numerical simulation.

The objection to this recommendation is easily anticipated: most biological roles are unseen in the fossil record, so direct tests are impossible. This is certainly true, but the same rationale applies to the behavioral context of most traits that would qualify as historical adaptation. The idea of variability selection implies an expanded importance of biological roles in hominins. There is no question that incorporating the concept into evolutionary morphology is currently a murky endeavor, but it is far worse to ignore this challenge and focus exclusively on linking traits with singular adaptations. That practice is relatively easy, but it is also misleading.

Neurobiological lability is one aspect of primate functional morphology that receives little attention, since it potentially impacts musculoskeletal function. Crompton et al. (2018) asserted that "neurobiological degeneracy"—the ability to carry out similar functions using dissimilar anatomical elements or configurations—deserves the same level of consideration for behavioral reconstruction as phenotypic plasticity. The problem, of course, is that this, too, will be largely invisible in the fossil record. The best we can probably do is to map out the scope of plasticity and degeneracy within and beyond phylogenetic brackets. A means of assessing the evolutionary likelihood of specified biological roles would be to articulate cost functions under different morphological conditions. This can be done via a purely thermodynamic criterion, provided that the mechanics of the biological role are modeled realistically. It might permit certain postulated functions to be rejected, most easily by assuming that performance versus integrity is a zero-sum game. This kind of approach has been articulated, but in a more generalized sense, as the evaluation of phenotype sets (Reeve and Sherman, 1993).

If I find myself deluded about the significance of some insight I have about what an aspect of morphology might mean in terms of adaptation, usually a consultation with a behavioral ecologist or field primatologist provides immediate remedy. It is not controversial to venture that there is a difference between what animals *can* do and what they *actually* do. In framing findings about the mechanical capacity of fossils, there will be contexts in which it is more prudent to specify what behaviors were truly out of bounds, rather than settle on a conclusion that selection targeted a particular function (Macho, 2013).

Virtually build the trait, following mechanobiological rules. If a reverse-engineered load case cannot produce the observed morphology, then the assumed mechanical context should be regarded skeptically. If the details of the load history are "incorrect" (i.e., the form produced is unobserved in nature), this calls for speculation that the trait developed in conjunction with unimagined functions. If the inferred function (the basis for the mechanical conditions) cannot produce the morphology, it might also mean that the trait in question is buffered against mechanical loads, or it may indicate a unique local mechanosensitivity.

This approach might be viewed as a perverse application of the paradigm method, fueled by a powerful CPU. But the paradigm method seeks a subjective agreement between an engineering solution and an evolved morphology, and it is not bound by any morphogenetic rules. A bone growth simulation is simply asking whether a hypothesized activity is capable of producing a particular range of morphology.

Gould and Lewontin's (1979) admonition resurfaces here, because the investigator may decide to tweak the load history iteratively, so the morphological fit improves. This aligns with these authors' derisive critique: "If one adaptive solution fails, invent another." Here again, however, morphogenetic rules have to be followed, any changes must be justified by some explicit rationale, and the numerical data remain subject to scrutiny. In any case, the alternative to testing another adaptive hypothesis is either (1) giving up or (2) invoking a nonadaptive hypothesis in its place. Nonadaptive status for traits is not objectionable in any way, but this status does not follow from the failure of one or more adaptive hypotheses.

They earn that status only if the trait's operation is identified and has no tangible contribution to fitness.

The articulation of phenotype sets can deflect Gould and Lewontin's (1979) complaint. The recommendation of Reeve and Sherman (1993) was that these sets constitute naturally occurring variants in a population or species, or they can be generated via theoretical models or analogies. The latter option must be governed by morphogenetic principles and have some empirical basis (e.g., expression in an analog or a variant created by an experimental manipulation). Otherwise, it is merely the paradigm method in disguise. The phenotype set is also anathema to the essentialist treatment of species-level traits, which is exactly why we should encourage its use.

Discourage invocation of constraint without identification. Morphologists (myself included) have always been able to extract themselves from a failed adaptive hypothesis by invoking constraint. Such statements are not usually followed by any suggestion of what the specific nature of the constraint is. This is the critical shortcoming of functional morphology, at least as it is practiced in both paleoanthropological and neontological contexts. We are now at the point where we can begin to articulate what these constraints are in the context of bone morphology (e.g., Cooper and Steppan, 2010; Hamrick, 2012; Percival et al., 2018).

Evaluate congruence through environmental modeling and ecological analogy. This is essential, yet operationally challenging. If the structuralist perspective on morphology is the correct one for understanding skeletal evolution, then hominins are not necessarily adapted to particular habitats. Instead, they are capable of finding environments that are suitable and variable over space and time. This requires a granular characterization of environment. Ecological analogs should not be cherry picked, and examples ought to be scrutinized via counterexamples.

Encourage publication of negative results. In Chapter 5, I alluded to the file-drawer problem as it pertains to functional morphology. There are a unknown number of well-designed comparative, experimental, or field studies that never see their way into journals because the proposed association between an environmental variable and a morphological one did not materialize. It is worthwhile to disseminate these findings as broadly as possible. The oft-heard rationale for ignoring such work is that, because no signal is found, there is no possible contribution to paleoanthropology. Negative findings can erode our confidence that our insights into hominin behavior are sharp and brilliant. Nonetheless, our goal—individually and collectively—is not to be regarded as right, but instead to be vigilant in determining if we really know what we think we know.

Chapter Seven

Teleonomy Revisited

> A scientist's first responsibility is skepticism toward his or her own most beloved conjectures.
>
> Fred Bookstein (2018:190)

Teleonomy does not exist as an established field in the biological sciences. That is not likely to change in the foreseeable future. The principal reason is that adaptation as an object of study separate from development and genetics (or phylogeny, in the context of paleoanthropology) is neither practical nor sensible. The lacunae that exist for examining the evolution of morphology are beginning to close. Though the genetic basis for morphology is still largely undiscovered, it appears inevitable that it will be worked out in great detail for several model species. What theoretical insights these advances will provide is hard to predict, but, at the very least, the morphogenetic factors underlying the discontinuities in morphospace will presumably come into clearer focus.

Functional morphology's most substantial advances will come from discoveries in behavioral ecology and experimental biology. Beyond identification of biological roles, it is critical to understand their lability from comparative musculoskeletal and neurophysiological perspectives (e.g., Nakajima et al., 2000; Vinyard et al., 2008; Menegaz and Ravosa, 2017; Granatosky et al., 2019). This endeavor will require a conceptual shift in emphasis from a narrow focus on discovery of historical adaptations to an examination of how anatomical complexes are recruited for multiple tasks (e.g., Nemeschkal et al., 1992). From this perspective, individual tasks (i.e., directed activity) represent biological roles. Any particular role will have certain performance requirements, and some of these may compromise the execution of other such roles. At the same time, there may be different ones that are relatively unhindered by these requirements. Figure 3.1 schematizes the problem and posits a general inattention to biological roles as an encumbrance to linking morphology to fitness. Functional morphological studies rarely explicitly consider the impact of organismal behavioral decisions, but such research foci are obviously crucial for understanding the emergence and maintenance of biological roles.

Experimental investigation into mechanical performance will remain an important component of hypothesis testing for behavioral and adaptive inferences. These approaches are indispensable for a full understanding of the functional capacity of anatomical complexes (Pigliucci, 2001). An experimental foundation is also essential for discovering phenotypic plasticity in the developing skeleton (Ravosa et al., 2007; Ravosa and Kane, 2017). The fact that plasticity may be largely indiscernible in the fossil record does not mean that we should be unconcerned with it in paleontological contexts, as plasticity was undoubtedly a consequential factor in human evolution. Neontological data on plasticity can only help in modeling its manifold effects in the fossil record. It would also be very useful to know how labile plasticity itself is across phylogenies; these data are certainly discoverable (Nunn, 2011; Kappeler and Fichtel, 2015).

It is unlikely that we will be able to recover the genomes of *Australopithecus*, *Paranthropus*, *Homo ergaster*, or any other early hominin as long as there is no organic phase recoverable from their remains. This vacuity might eventually become the greatest obstacle to a fully coherent understanding of hominin functional morphology. We can circumscribe the genomes of Pliocene hominins from ourselves, Denisovans, Neanderthals, and chimpanzees, but we have no way of knowing patterns of variation *within* those taxa for which the genetic data are inaccessible. Our best hope is that hominin phylogeny can be brought to bear on questions of character transformations and their possible significance. The problem remains, of course, that the same morphological data used to make taxonomic decisions also serve as the basis for adaptive and behavioral inferences. The morphospecies concept is therefore no friend of the functional morphologist.

Behavioral data from living primates are plentiful, but paleoanthropologists generally recruit these as post hoc support for hypotheses. As Bock (1980) and others since then have noted, to apply behavioral data productively to functional morphology, definitions of variables and field strategies for collecting the relevant data need to be worked out in advance. Otherwise, behavioral data originally recruited to address a neontological problem may be applied to a paleontological context in which different questions are being asked.

E. Russell's (1916) argument that behavior is the operationalization of morphology was widely panned, not because the idea was wrong, but because of the way the sentiment was articulated. He saw the organism—the individual—as autonomous, able to operate in some fashion independently of history, thereby justifying the charges of mysticism and Lamarckian reasoning (Roll-Hansen, 1984). Part of the problem we have in seeing behavior as something durable and potent, relative to morphological evolution, is that we experience it as ephemeral or unstable, and therefore undurable. We unconsciously think of it as elective, determined solely by immediate circumstances. Yet Greene (1994) notes that behavior simply operates on a temporal scale intermediate between relatively instantaneous biochemical processes and apparently "unchanging" morphology. The temporal scale of change may seem distinct, but all of these processes are subject to natural selection.

Behavioral study is quite obviously the way in which we recognize biological roles, and it is this research area that will illuminate how important these roles are in determining fitness over short and long terms. If nothing else, attention to biological roles will serve to move paleoanthropology away from essentialist tendencies when drawing conclusions about the significance of skeletal traits in the fossil record. It is difficult to anticipate how behavioral variation impacts skeletal differences, but reductively modeling behavior as stress stimuli and investigating their morphological effects on bone is one means of clarifying these impacts. Equifinality may be indicated; this will be unwelcome, but we will then know what kinds of speculations are valid.

The theoretical principle that bone mechanics represent the management of performance, balanced against structural integrity, must generally be true, given the operation of natural selection, but important questions remain to be fully worked out. Safety factors in bone are probably not homogeneous within or across skeletal elements or across species (Ravosa et al., 2000; Skedros et al., 2003; Blob et al., 2014). Is this finding explicable on mechanobiological or physiological grounds, or have we conceptualized safety factors inadequately? In Chapter 2, I speculated that fatigue resistance is as likely a target of selection as yield strength, but that argument does not yet have the appropriate data for confirmation.

Conjuring Human Evolution with Numbers and Skepticism

As I noted in Chapter 2, theoretical morphology is a recognized but largely unapplied perspective for paleoanthropological research. To deny that *Paranthropus* had limited morphological options for dental and facial evolution is a functional-

ist position. Genetic and developmental data should convince us that a structuralist perspective is better equipped to explain apparent mismatches of ecology and
morphology (Ungar and Hlusko, 2016). Simulation models are vital, not because
they allow correct inference, but because they delineate tangible possibilities from
those that are analytically or processually inconceivable. Such models are probably more valuable when applied post hoc rather than a priori, if only because,
without existing data, the appropriate scope of parameters may be largely guess-
work. Post hoc models have the benefit of real-world data to which the output
should adhere.

Functional inference becomes most persuasive when the processes underlying
morphological variation and innovation are identified. The paleontological record
does not permit direct observation of process, and nothing less than a time machine will change this. Consequently, there is a high degree of uncertainty inherent in the enterprise. This recognition is not banal if it inspires a culture of practice that is dedicated to understanding the depth and scope of that uncertainty.
Equifinality is a "problem" only insofar as the processes underlying it remain mysterious. Theoretical morphology, particularly when it utilizes developmental and
mechanobiological perspectives, promises to demonstrate how a finite morphospace is compelled by evolutionary processes.

Mechanobiology is not a panacea for all the challenges of functional inference
in paleoanthropology. It is a complement to, rather than a replacement for, ecological modeling of human evolution. Its promise lies in the idea that, as far as
bone is concerned, the environment can be distilled into a set of physiological conditions. There is noise in the system, but that should not bother us, given the
nature of biological data. The fact that bone morphology does not match our model
of the world is not a failure of bone function; rather, it is an indictment of our
model. We should take mechanobiology as far as it will go, until it has nothing left
to offer the paleontologist. We will discover regularities in bone development, unanticipated phylogenetic variants, and explanations for why some adaptive solutions just cannot happen. The mechanobiological perspective has already allowed
us to throw off the yoke of Wolff's law and carve out validated aspects of the Mechanostat and incorporate them into new models.

To some degree, the foregoing discussion provides a remedy for the difficulties
inherent to functional inference in human evolution that I articulated in the open-
ing chapter. These issues can be restated as follows: (1) there are some problems
that we do not have the means to address, because the needed data are invisible;
(2) the identification of historical adaptation provides a limited window into

behavior; and (3) hypotheses that purport to offer the broadest explanations are the most resistant to refutation. These observations are neither novel nor, I assume, controversial. I have argued plainly for a commitment to reductionist practice in paleoanthropology, because I believe this approach leads to more consistently reliable knowledge. Philosophically, reductionism can be fairly criticized as flawed, because inherent in its goal to explain complex phenomena simply is the shortcoming of delineating them incompletely. Methodological reductionism, however, has the potential to address the persistent challenges in paleontological inference, as long as we accept that settling for incremental advances in knowledge, as opposed to seeking sweeping explanations for human evolution, is worthwhile.

Bad Paradigm or Bad Practice?

Commenting on paleoanthropological praxis, Chamberlain and Hartwig (1999:43) maintained that, "as a collection of scholars, we have the curious habit of adhering to a knowledge claim we arrived at through Kuhnian science while at the same time disavowing the possibility that knowledge can be acquired through anything but a positivist approach." Their position was that positivism is impotent in resolving disputes within the discipline (exemplified by those outlined in Chapters 4 and 5), because paradigmatic dissensions preclude accommodation to conflicts of interpretation. This diagnosis is defensible from a philosophical perspective, but the pragmatic implications are discouraging. If, nominally, all paradigms have validity, and data sets are best interpreted from a particular paradigmatic focus, then any one scenario of human evolution is as good as any other.

While a positivist framework for conducting research could be characterized as naïve or inefficient, it has been beneficial to paleoanthropology over the past half century. We have more to learn about fossil bones from the experimental literature in skeletal biology than we do from comparing the consequences of punctuated equilibria and phyletic gradualism for the interpretation of morphology (e.g., Tattersall, 2000). Paleoanthropology has never suffered from a dearth of ideas. The challenge, rather, has always been to produce data to simultaneously promulgate *and* constrain hypotheses.

In their essay, Chamberlain and Hartwig (1999) insisted that it is critical to understand what they labeled "Kuhnian" dynamics in paleoanthropology. One presumes, from their argument, that the paradigm does not differ from evolutionary biology to any significant degree, even if the ideas diffuse rather slowly from mainstream zoology to anthropology. The epistemological tension within paleoanthropology arises between the positions that hominins are *sui generis*—comprehensible only on their own terms—and that hominins represent an outlier among Animalia, albeit one that can only be understood in the context of biodiversity (Marks, 2005). This conflict between perspectives can boil down to basic arguments, such as whether the anatomy of monkeys can tell us anything important about the particulars of hominin fossil morphology. The ensuing judgments create paradigmatic filters with respect to what gets published and where, and the impact of this on disciplinary content can be substantial.

In the specific context of functional inference in paleoanthropology, the problem is one of a mismatch between theoretical commitments and methodological conduct. Though principles of structuralism can serve as theoretical guides, research has historically regressed to functionalist explanations. This is beginning to change, however, as investigations go beyond pattern recognition to inferences of underlying process (Nunn, 2011). To our credit, we do not reject advances in genomics or developmental biology simply because such things are invisible in the fossil record. Nonetheless, the mapping of structure to function is a good deal more nuanced than what we understood it to be only 30 years ago. Despite tremendous advances in morphometrics and the virtual and numerical distillation of form, our improved characterization of fossils—especially who they resemble and how they resemble them—will always be insufficient, in itself, for reliable behavioral inferences.

Lowering Expectations Now for a Mature Science Later

If we could momentarily have omniscience to scrutinize functional-adaptive hypotheses in human evolution and characterize them as either factually supported or refuted, I expect most paleoanthropologists would wager the number in the latter category would be larger, by a wide margin. That pessimism is justified for several reasons, including the methodological bias to functionalist interpretation. Operationalizing a structuralist methodology will require collaboration and conversation on a grand scale among behavioral ecologists, paleontologists, geneticists, and developmental biologists (White, 2000) and will, at least for the time being, probably render a judgment that some questions of behavioral inference are unanswerable. This is something we should want to know. Uncertainty is not our enemy. We will do better at understanding our origins by valuing both epistemological humility and pluralistic methodology.

Notes

CHAPTER ONE: **Unresolved Problems in Human Evolution**

1. Except when the fossils are teeth, trackways, preserved soft tissues, or coprolites. My focus in this book, however, is squarely on bone morphology and evolution. Teeth are not bones, do not develop in the same manner as bones, and, in any case, have been studied to death—to our collective benefit—in the hominin fossil record. Teeth will feature prominently in Chapter 5. In the meantime, fossils are old bones, unless otherwise noted.

2. Speculation as to what inferences will forever be impossible to draw is not recommended, per Clarke's rule (coined by the science fiction writer Arthur C. Clarke in 1972): "When a distinguished but elderly scientist states that something is possible, he is almost certainly right. When he states that something is impossible, he is very probably wrong."

3. Should readers make it through this book, they will note that the treatment of the hominin fossil record is very uneven, with most attention given to australopiths (Chapters 4 and 5) and a brief foray into later Pleistocene *Homo* (Chapter 6). The principal reasons for this bias in coverage are first, I am more familiar with australopith hypodigms than those of early *Homo*, and second, *Homo* presents larger problems for functional inference, because of our enhanced cultural workarounds to ecological problems. In effect, I am chickening out of the more challenging fossil inventory as far as behavioral reconstructions go. Instead, my goal in this book is to advocate for a complementary approach to functional inference, rather than to crack the deeper mysteries of human emergence. If this path cannot bear fruit for australopiths, it is probably also hopeless in the case of *Homo*.

CHAPTER TWO: **Situating Functional Morphology in Evolutionary Biology**

1. Pleiotropy describes the action of a single gene in which two seemingly unrelated phenotypic traits are produced. Its reality has profound implications for morphological evolution.

2. The beam concept is borrowed from engineering and is widely applied in biological anthropology and comparative anatomy; manufactured examples include bridges, diving boards, and two-by-fours. By definition, beams are bent. This incompletely describes deformations of the supraorbital torus. The principal theoretical objection is that it is firmly anchored to a facial skeleton rostrally and to a calvarium posteriorly. It cannot possibly be loaded with the elegant simplicity described in strength-of-materials texts.

3. The critical stress in Biewener's (1993) formulation was yield stress; i.e., the stress at which permanent deformation sets in after removal of load. Under yield stress, bone may maintain its gross size and shape, but it is damaged nonetheless and has begun its journey to failure. Another way of calculating a safety factor would be the division of ultimate stress by physiological stress. Under ultimate stress, one sees structural fracture.

4. You have applied the principle of fatigue if you have ever had to contend with a defective can opener. This device leaves part of the lid circumference uncut, so then you have to grab a free margin of the lid and wiggle it back and forth. The resistance of the intact remnant of the lid declines quickly, until the can's contents become fully accessible. The number of wiggles it takes to liberate the lid is its fatigue life.

5. Strain refers to deformation, and it is conventionally defined as the change in length of an object over its original length. Strain can be thought of as a manifestation of stress.

6. Ultimate causation in structuralist morphology was conceived of by 19th-century transcendentalist anatomists in terms of archetypes. These were not ancestors, but instead were ideal or pure morphological configurations from which groups of living forms (e.g., vertebrates) were derived. The object of transcendentalist study was the identification of homology, but without an explicit evolutionary connotation. Phylogeny represents ultimate causation in constructional morphology.

7. These are necessary effects of scale on physiology. Compared with a tarsier-sized fossil, we can be very confident that a gorilla-sized one had a lower heart rate, weighed more, and could not afford to fall out of a tall tree.

8. This should not be confused with Haldane's dilemma or Haldane's rule. Haldane's paradox is defined in so many different ways across the literature that it is clear people are not always talking about the same problem. I am using it in the sense that Levinton (2001) intended, and no other.

9. The evidence that bone material stiffness changes systematically with body size is sparse; Erickson et al.'s (2002) comparative study indicated relatively invariant material stiffness in the vertebrate femur.

10. Noise in this context is unexplained variance; i.e., deviation of an observed value from that predicted by the bivariate coefficients.

11. Scaling across a restricted size range of taxa represents a narrow allometric approach. Its advantage is that subtle scaling differences that would go undetected in a broad (e.g., shrew-to-elephant) survey can be quantified. This can be effective for evaluating patterns of morphological variation within a focal clade, especially in deciding whether ecological factors or physiological imperatives account for observed patterns.

12. Perhaps predictably, the principal arguments surrounding *Au. sediba* and *H. floresiensis* are taxonomic/phylogenetic (Argue et al., 2017; Kimbel and Rak, 2017). These assertions are obviously based on morphology, with affinities to *Homo* versus australopiths being focal.

13. On deeper analysis, this may turn out not to be true. Homoplasies may mean that character evolution is bumping up against some sort of morphogenetic limit.

Specific examples with respect to the primate skeleton are hard to demonstrate, but the status of a trait as homoplasy or homology is always relative to some other taxon, so as far as the organism is concerned, that *status* is immaterial, as opposed to the trait's *utility*.

14. Principal strains, in this context, can be thought of as tension and compression, which are acting at right angles to one another. This happens whether a bone is being bent, twisted, sheared, or axially loaded.

15. At this point, the drudgery of formal definition can no longer be avoided. As noted before, stress is defined as force acting over an area, a sort of standardized internal load; it can be thought of colloquially as pressure. Strain is the deformation of material in association with a load, reckoned as a proportional change in length or as an angular distortion. Strain is dimensionless. The SI standard for pressure is the pascal; one newton of force (N) distributed over a square meter. The magnitude of a newton can be appreciated by rolling an apple off a tabletop, shortly after which it will impart a force of 1N (more or less) on the floor, depending on the size of the fruit and the height of the table. Stresses in bone are usually reported in megapascals (MPa) or gigapascals (GPa), or a million and a billion pascals, respectively.

16. Anisotropy means "not isotropic," or having different behaviors when measured in different directions. In the vernacular, the word refers to something that has a grain: for example, wood is anisotropic. For trabecular bone, a lattice that is not symmetric in all dimensions will qualify as anisotropic.

17. Volume fraction refers to the ratio of the volume of bone tissue to the total volume of the sample. In the case of a cube of trabecular bone, the volume of all the pores plus all the bone tissue is the total sample volume.

18. A microstrain is one millionth of a strain. Since strain is calculated as the change in length over original length, one $\mu\varepsilon$ is a change of 0.0001%.

19. In much of the anthropological literature, all bone activity (apposition, resorption, repair) is described as a process of remodeling. In the skeletal biology literature, remodeling only refers to a specific sequence of bone cellular activity, involving resorption of a bone surface, followed by a nearly equivalent replacement of material. Modeling refers to the addition and resorption of bone without site-specific, coordinated sequential activity of osteoclasts and osteoblasts. Growth, geometric changes in bone structure, and cortical drift are all examples of modeling. I follow the skeletal biology convention throughout. Consequently, remodeling is associated with localized cellular activity on intracortical and trabecular surfaces that does not result in gross structural changes in size, shape, or spatial position.

20. This risk in determined by the interaction of strain magnitude and the number of accumulated loading events. Cycles to failure will be less if strains are higher.

21. In the process of bone modeling and remodeling, some of the cells that lay down bone matrix (osteoblasts) end up getting buried in this matrix. Once this happens, they are renamed osteocytes. Their cell bodies reside in lacunae within the bone matrix and have cytoplasmic/dendritic extensions that occupy tiny tunnels (canaliculi) that radiate from the lacuna. The arrangement facilitates communication among neighboring osteocytes via gap junctions. This network of intercellular communication plays a

vital role in mechanotransduction of tissue strains and, ultimately, bone adaptation in the broad sense.

22. By definition, beams are loaded perpendicular to their long axis, and columns are loaded parallel to that axis. Columns therefore always experience axial loads, while beams may or may not. When bones are described as beams or columns, this means that bending is considered to be the important loading mode.

23. A state of stress or strain in a finite volume cannot be completely described by a single number or a single vector. The determination of a stress or strain tensor is a means of enabling that description. If one envisions a cube of bone material under load, it can be shown that there are six mutually perpendicular elements of stress to consider: three "normal" (perpendicular to the cube faces), and three "shear" (acting parallel to the cube surfaces). If that were not nightmarish enough, if you redefine your perspective (mathematically speaking, change your coordinate system), all of these stress values change.

CHAPTER THREE: **Approaches to Functional Inference in Paleoanthropology**

1. It is becoming less typical to see categorical descriptions of morphology, because the technologies for gathering size and shape information are widely available. On the other hand, the phylogenetic significance of morphological characters is facilitated by conceptualizing them as discrete. This, of course, reinforces essentialist ideas, and these still bleed over into functional investigations.

2. This qualification is needed, because lab-based experimental investigations, which can measure performance with great precision, may not, by themselves, tell us anything about biological roles.

3. In morphometrics, shape does not exactly mean geometry, but it is supposed to be reflected in variables for which the effects of size have been removed, or at least accounted for. This is a fully justified statistical procedure, rather than an actual removal of size effects from a biological comparison.

4. Selection—by definition—destroys variation, and Potts (1996, 1998) went to considerable lengths to assure readers that he was not arguing for natural selection doing the opposite.

5. Eurytopic (loosely translated as "broad place") species are tolerant of a range of environments; stenotopic (steno = narrow) taxa are adapted to specific environments.

6. Owen Lovejoy, Bruce Latimer, James Ohman, and their collaborators disagree. Chapter 4 explains why.

7. Beanbag, in this context, refers to the conceptualization of alternative alleles as different colored beans, which provide material for laboratory exercises to illustrate Mendelian genetics. Mayr (1963) questioned the sufficiency of this metaphor to explain organismal evolution. While this criticism continues to be challenged, the reality of gene interactions and epigenetic sources of variation (with respect to morphology) means that it is generally not useful to think of single-locus determinants of anatomical traits; i.e., we do not expect to find a gene for a valgus knee.

8. The exception would be if nobody read the paper, or people read it but decided it was uninformative. Consequently, having to defend a model is a good thing.

9. Geometry refers to generalizations about shape. FEA, for example, is not restricted by certain ranges for span:depth ratios and prismatic shape, which are required for beam theory to generate correct predictions. Homogeneity refers to the invariance of structural and material components. The intrinsic variation in bone density, most clearly exemplified by looking at cancellous versus cortical bone, means that it is not homogeneous. Material properties (e.g., stiffness) are allowed to vary in FEA models.

10. This is a fairly liberal criterion. In a normal distribution, two standard deviations encompass about 95% of the expected values. By this scheme, given certain distributions of experimental data, model strain values that are double or half of the average experimental values may be deemed valid.

11. Strait et al. (2009:Supporting Information, 3) denied that this decision had any meaningful consequences: "The omission of the fascia does not appear to affect strains outside of the arch, and thus does not threaten the integrity of the model as a whole." The logic of this statement is that some of the predicted strains elsewhere in the model are better, and it presupposes that the effects of the fascia are exclusively local. Kupczik et al. (2007) suggested this is an overgeneralization.

12. This refers to an unpublished model in which environmental change happens more or less instantaneously. Interested readers are welcome to review a very inefficient R script by request to the author.

13. The problem was that the ratio used to calculate relative size did not use dimensionally equivalent variables; i.e., relative size was determined as an area over a volume. Since volume increases faster than area in larger animals (mathematically obligate for similar geometries), lemurs had very small denominators in the ratio and, thus, larger values of relative size.

CHAPTER FOUR: **Bipedality**

1. The core group of investigators in this camp hold or have held positions at institutions in Ohio, including the Cleveland Museum of Natural History, Case Western Reserve University, and Kent State University. The extended cohort of collaborators are based at institutions world wide. Not too much significance should be attached to the efficiency moniker. This suggests economy, which, in turn, has energetic implications, but a developed argument about the details of metabolic costs was not central to the initial debate, even though such considerations were relevant. If the reader prefers, the "efficiency" descriptor could be renamed "equivalency."

2. Jack Stern and Randall Susman, in addition to their frequent collaborators Bill Jungers and Brigitte Demes, held positions at what is now called Stony Brook University in New York State. As a matter of full disclosure, my doctorate was conferred there, and Stern and Jungers were members of my supervisory committee. While my dissertation had nothing to do with locomotion, I have no doubt that my thinking on matters related to the evolution of bipedality was influenced by my choice of graduate programs.

3. The reader can become convinced of this by spending several uninterrupted minutes walking, using a compliant gait. Just keep your knees bent and your trunk leaning forward. Groucho Marx provides a good model.

4. An intermembral index is the ratio of the sum of humerus and radius lengths over the sum of femur and tibia lengths. By convention, this ratio is multiplied by 100 and rounded to a whole number.

5. Stern and Susman (1983) defined this variable as the angle formed by the ventral bar of the subscapular fossa and a line joining the superior and inferior "limits" of the glenoid cavity. This variable was chosen over one utilizing the axillary border, on the grounds that the ventral bar is medial to and not coincident with the axillary border, which is subject to a good deal of idiosyncratic variation and presents no clear landmarks for standardizing its orientation. Despite the apparent potential for considerable measurement error for this variable, there is remarkable agreement among otherwise dissenting investigators as to its value for AL 288-1.

6. This allometric correction is crucial for Lucy but less problematic for *Au. afarensis* as a whole. The species was dimorphic, and males may have been anywhere from half again to nearly twice as large as females (McHenry, 1992). KSD-VP-1/1g, the scapula of which is discussed subsequently, is assumed to be male by virtue of its large size (Haile-Selassie et al., 2010).

7. Often in the literature, this reinforcement is expressed in terms of stress; i.e., "bone is thickest where stress is highest." This is misleading if taken literally. The general principle is that bone mass is regulated to maintain stress *within certain limits*. Thus differences in bone area or mass do not reflect different stress magnitudes; they indicate broadly similar stresses.

CHAPTER FIVE: **Hominin Dietary Adaptations**

1. This is a mammal-centric perspective. Alligators do not have molars, nor do they chew their food, as mastication is not imperative for them. The metabolic argument that chewing is necessary for digestive throughput in mammals is nonetheless sound, as long as we ignore cetaceans and edentates.

2. Kay (1975) examined metabolic scaling in primates with reference to molar size. He calculated an expected exponent to which primates should adhere in order to meet energetic requirements. To his surprise, primate teeth do not meet this benchmark. He sided with empiricism, rather than theory, by noting that since large primates were not starving to death as a result, the assumptions of metabolic scaling needed rethinking.

3. The reason for this peculiar list will be made apparent in the discussion of isotopes that follows.

4. Grade refers to an often nebulously defined degree or level of adaptive organization. Huxley (1958) distinguished the concept from clades, which, by definition, are constrained to monophyletic units. Following the widespread adoption of cladistic methods in biological anthropology, the term is often used pejoratively. Nevertheless, if the concept were operationalized, it could have value in comparative applications.

5. See C. F. Ross et al. (2012) for a review. There are dozens of papers exploring the covariance of diet with aspects of mandibular morphology. In several cases, a null hypothesis—i.e., that there is no predictable covariance—is credible. It is instructive that explanations for the lack of correspondence are rarely attributed to the possibility that there is no relationship. Instead, variable choice or poor dietary specificity is blamed.

6. This ignores the real possibility that overall skull architecture can predispose the temporomandibular joint to high distractive forces, which would encourage an animal to regulate its maximum biting effort. Ledogar et al. (2016) suggested this may have constrained masticatory function in *Australopithecus sediba*.

7. Only one living primate, the tarsier, is a committed carnivore. Being very small, nocturnal, and remarkably strange anatomically, it has not been offered up as an instructive analogy for meat-eating in hominin evolution.

8. Grine (1981) rejected both Szalay's (1975) and Cachel's (1975) analogies, on the grounds that hominin teeth and carnivore teeth were poor functional analogs, based on their distinctiveness.

9. In this instance, I am excluding the pathological context. Bone hypertrophy can occur in the absence of altered physical activity, but it is convenient, and hopefully prudent, to assume that this is not what we are seeing most of the time in the fossil record.

10. The rapid turnover of microwear should not be terribly surprising, as enamel experiences brittle fracture. One bite can introduce new wear features and partially obliterate others. Grine (1986) referred to this as the "Last Supper" effect.

11. An alternative explanation of the difference in *Paranthropus* microwear fabrics is that the East African form was more fastidious in its food preparation. There is no direct evidence for this, however, from the fossil record.

12. This concept does not assume that no nuts or seeds are found in open-country habitats. Primate seed-eaters overwhelmingly harvest from C_3 species. An important exception is *Theropithecus*, which served as Jolly's (1970) hominin analog for the seed-eating hypothesis. The grass seeds consumed by geladas do not produce microwear fabrics similar to those seen in mangabeys or tufted capuchins.

13. Strait et al. (2008, 2013) were correct, of course, in saying that what one actually eats is not necessarily an indicator of what one is adapted to eat (for example, offer a doughnut to a dog). But for both *Au. africanus* and *P. boisei*, they were effectively arguing that the microwear and isotopic data should be reinterpreted or else ignored, given contrary biomechanical interpretations.

14. The identity of the ancestor among australopiths is not important to this argument. As noted before, megadontia, thick enamel, and buttressed jaws are present early in the radiation. The advent of facial retraction is more uncertain, given *Au. aethiopicus*.

15. The details of model critiques and arguments over consilience are covered in Grine et al. (2010), Strait et al. (2012, 2013), and Daegling et al. (2013) and are not rehashed here. Despite the invective undercurrents in these works, the light:heat ratio is not zero.

CHAPTER SIX: **The Osteocyte Perspective on Human Evolution**

1. Physiological adaptation is sometimes referred to as "functional adaptation." Since historical adaptation is also routinely conceived of in functional terms, the term "physiological adaptation" is preferred, as it communicates the idea of change that occurs during the lifespan.

2. Growth (i.e., changes in size) of bone can be described as modeling, but the distinction here is useful, because somatic growth of the skeleton, to some degree, occurs independently of the vagaries of daily load stimuli. One can grow up running two miles a day or watching television instead and, in both cases, arrive at adulthood with a competent skeleton (although differences in its quality are probably present). In anticipation of the objection regarding the phenomenon of bone loss in astronauts, the germane experiment would be one that follows an animal model through ontogeny in zero gravity. One striking discovery in astronauts and the mice accompanying them in space flight is that non–weight-bearing bones (e.g., those of the skull) do not experience the magnitude of bone loss seen in the femur and lumbar spine (Tavella et al., 2012).

3. Because normal (longitudinal) strains usually vary in magnitude from location to location, shear strains are compulsory to fulfill thermodynamic equilibrium requirements. The one exception to this is the phenomenon of "pure" bending, in which the bending moments and their accompanying normal strains are constant along a beam. There is no experimental evidence to indicate that this ever happens in living bones.

4. What Carrier and Morgan (2015) assumed was that a tough diet posed no skeletal challenges, whereas a hard-object diet did. Tougher diets still involve appreciable strains but, more importantly, require more daily work, which increases fatigue risk in bone.

5. Attempts have been made to argue that other animals have chins (e.g., elephants), but arguments for the homology of these resemblances have not provided much convincing developmental data.

6. For historical reviews of hypotheses concerning the evolution of the chin, see Du Brul and Sicher (1954), Daegling (1993), Schwartz and Tattersall (2000), and Pampush and Daegling (2016).

7. This supposition is consistent with *in vivo* strain data (Hylander, 1984), except that to describe the load as "bending" is a misapplication of beam theory. As was the case with the femoral neck (see Chapter 4), a bending moment acting here produces a great deal of shear, which is even more threatening to bone integrity than tension. In this case, the mechanobiological perspective could have been informative, but the mechanical model for interpreting the morphology was wrong.

8. Concentric cross-sections are optimized for torsional shear strains. Under these initial conditions, the geometry of the bone will not alter, provided the bone growth algorithm is rational. Nowlan and Prendergast (2005) were exploring how changes in cortical thickness, not geometry, evolved.

9. Bone cells, strictly speaking, are equipped to sense strain, rather than stress per se. Throughout the discussion, "strain" can be substituted for "stress" without requiring a change in interpretation. This, however, requires the relationship of stress to strain to be more or less constant. In fact, bone tissue varies spatially in this parameter (elastic modulus) whenever investigators have looked into it (Rapoff et al., 2008). Whether this introduces a systematic bias is unclear (Le et al., 2017).

10. This model was originally presented by the author at the 2019 Paleoanthropology Society meetings in Albuquerque, New Mexico. It has been modified since then to improve the calculation of the stress field and more realistically model apposition and resorption patterns (Daegling et al., 2021).

11. No distinction is made here between a daily stress stimulus and an effective stress stimulus, because (1) the assumption is that the loading profile is homogenous; (2) the load is "averaged" across a week; and, most importantly, (3) the goal of the exercise is to ascertain whether distinct mechanical responses will "grow" different bones, which requires comparable loading regimens across contexts. An effective stimulus can be incorporated into such models without too much difficulty, but the added computational time is not insignificant. Nowlan and Prendergast (2005) provided a framework for turning this purely developmental simulation into an evolutionary one.

12. Conceiving of an isostress condition in a three-dimensional (i.e., real) bone is even less credulous. It must also be borne in mind that bones encounter multiple load cases daily. The probability that all of these are characterized by isostress conditions is roughly zero.

13. The noise incorporated into load magnitude on a weekly basis does not count as a nuisance variable, because the algorithms were all quite robust regarding this stochastic factor; i.e., repeated runs with idiosyncratic load histories had no meaningful impact on results. Potential nuisance variables—such as hormonal or nutritional disruptions, signaling errors, or compromised cellular activity—are quite easily incorporated into this type of model, once a decision is made about what the proximate effects of these are in terms of apposition and resorption rates.

14. One could take issue with this by noting that chest pounding would produce a stress stimulus on the ribs that theoretically would be detectable by examining osteon density or costal hypertrophy. My critique of Carrier and Morgan (2015), however, also applies in this case. Ribs are heavily remodeled in people; thus a chest-pounding osteonal signature is unlikely to be detected.

References

Ackermann, Rebecca Rogers, and James M. Cheverud. "Detecting genetic drift versus selection in human evolution." *Proceedings of the National Academy of Sciences* 101, no. 52 (2004): 17946–17951.

Adams, Neil F., Thomas Gray, and Mark A. Purnell. "Dietary signals in dental microwear of predatory small mammals appear unaffected by extremes in environmental abrasive load." *Palaeogeography, Palaeoclimatology, Palaeoecology* 558 (2020): 109929.

Ahern, James C. M. "Foramen magnum position variation in *Pan troglodytes*, Plio-Pleistocene hominids, and recent *Homo sapiens*: Implications for recognizing the earliest hominids." *American Journal of Physical Anthropology* 127, no. 3 (2005): 267–276.

Alberch, Pere. "Ontogenesis and morphological diversification." *American Zoologist* 20, no. 4 (1980): 653–667.

Alexander, R. McN. "The ideal and the feasible: Physical constraints on evolution." *Biological Journal of the Linnean Society* 26, no. 4 (1985): 345–358.

Alexander, R. McN. "The scope and aims of functional and ecological morphology." *Netherlands Journal of Zoology* 38, no. 1 (1987): 3–22.

Alexander, R. McN, A. S. Jayes, G. M. O. Maloiy, and E. M. Wathuta. "Allometry of the limb bones of mammals from shrews (*Sorex*) to elephant (*Loxodonta*)." *Journal of Zoology* 189, no. 3 (1979): 305–314.

Anapol, Fred, Nazima Shahnoor, and Callum F. Ross. "Scaling of reduced physiologic cross-sectional area in primate muscles of mastication." In *Primate Craniofacial Function and Biology*, edited by Christopher Vinyard, Matthew J. Ravosa, and Christine E. Wall, 201–216. Boston: Springer, 2008.

Anderson, Philip S. L., Jen A. Bright, Pamela G. Gill, Colin Palmer, and Emily J. Rayfield. "Models in palaeontological functional analysis." *Biology Letters* 8, no. 1 (2012): 119–122.

Arcadi, A. Clark. "Species resilience in Pleistocene hominids that traveled far and ate widely: An analogy to the wolf-like canids." *Journal of Human Evolution* 51, no. 4 (2006): 383–394.

Ardey, Robert. *The Hunting Hypothesis: A Personal Conclusion Concerning the Evolutionary Nature of Man*. New York: Macmillan, 1976.

Argue, Debbie, Colin P. Groves, Michael S. Y. Lee, and William L. Jungers. "The affinities of *Homo floresiensis* based on phylogenetic analyses of cranial, dental, and postcranial characters." *Journal of Human Evolution* 107 (2017): 107–133.

Ariew, André. "Ernst Mayr's 'ultimate/proximate' distinction reconsidered and reconstructed." *Biology and Philosophy* 18, no. 4 (2003): 553–565.

Arnold, Patrick, Martin S. Fischer, and John A. Nyakatura. "Soft tissue influence on *ex vivo* mobility in the hip of *Iguana*: Comparison with *in vivo* movement and its bearing on joint motion of fossil sprawling tetrapods." *Journal of Anatomy* 225, no. 1 (2014): 31–41.

Arnold, Stevan J. "Morphology, performance and fitness." *American Zoologist* 23, no. 2 (1983): 347–361.

Arnold, Stevan J. "Constraints on phenotypic evolution." *American Naturalist* 140 (1992): S85–S107.

Atchley, William R., and Brian K. Hall. "A model for development and evolution of complex morphological structures." *Biological Reviews* 66, no. 2 (1991): 101–157.

Auletta, Gennaro. "Teleonomy: The feedback circuit involving information and thermodynamic processes." *Journal of Modern Physics* 2, no. 3 (2011): 136–145.

Baab, Karen L., Lynn E. Copes, Devin L. Ward, Nora Wells, and Frederick E. Grine. "Using modern human cortical bone distribution to test the systemic robusticity hypothesis." *Journal of Human Evolution* 119 (2018): 64–82.

Backwell, Lucinda R., and Francesco d'Errico. "Evidence of termite foraging by Swartkrans early hominids." *Proceedings of the National Academy of Sciences* 98, no. 4 (2001): 1358–1363.

Bailey, Katie E., Susan E. Lad, and James D. Pampush. "Functional morphology of the douc langur (*Pygathrix* spp.) scapula." *American Journal of Primatology* 79, no. 6 (2017): e22646.

Barak, Meir M., Daniel E. Lieberman, and Jean-Jacques Hublin. "A Wolff in sheep's clothing: Trabecular bone adaptation in response to changes in joint loading orientation." *Bone* 49, no. 6 (2011): 1141–1151.

Barel, C. D. N. "Concepts of an architectonic approach to transformation morphology." *Acta Biotheoretica* 41, no. 4 (1993): 345–381.

Bastir, Markus. "Back to basics: Morphological analysis in paleoanthropology." In *Rethinking Human Evolution*, edited by Jeffrey H. Schwartz, 205–227. Vienna Series in Theoretical Biology No. 21. Cambridge, MA: MIT Press, 2018.

Beaupré, G. S., T. E. Orr, and D. R. Carter. "An approach for time-dependent bone modeling and remodeling—theoretical development." *Journal of Orthopaedic Research* 8, no. 5 (1990): 651–661.

Bedaso, Zelalem K., Jonathan G. Wynn, Zeresenay Alemseged, and Denis Geraads. "Dietary and paleoenvironmental reconstruction using stable isotopes of herbivore tooth enamel from middle Pliocene Dikika, Ethiopia: Implication for *Australopithecus afarensis* habitat and food resources." *Journal of Human Evolution* 64, no. 1 (2013): 21–38.

Behrensmeyer, Anna K., and Kaye E. Reed. "Reconstructing the habitats of *Australopithecus*: Paleoenvironments, site taphonomy, and faunas." In *The Paleobiology of Australopithecus*, edited by Kaye E. Reed, John G. Fleagle, and Richard E. Leakey, 41–60. Dordrecht, Netherlands: Springer, 2013.

Bentolila, V., T. M. Boyce, D. P. Fyhrie, R. Drumb, T. M. Skerry, and M. B. Schaffler. "Intracortical remodeling in adult rat long bones after fatigue loading." *Bone* 23, no. 3 (1998): 275–281.

Bergmann, G., G. Deuretzbacher, M. Heller, F. Graichen, A. Rohlmann, J. Strauss, and G. N. Duda. "Hip contact forces and gait patterns from routine activities." *Journal of Biomechanics* 34, no. 7 (2001): 859–871.

Berillon, Gilles. "In what manner did they walk on two legs?" In *From Biped to Strider: The Emergence of Modern Human Walking, Running, and Resource Transport*, edited by D. Jeffrey Meldrum and Charles Hilton, 85–100. Boston: Springer, 2004.

Berillon, Gilles, Guillaume Daver, Kristiaan D'Août, Guillaume Nicolas, Bénédicte de la Villetanet, Franck Multon, Georges Digrandi, and Guy Dubreuil. "Bipedal versus quadrupedal hind limb and foot kinematics in a captive sample of *Papio anubis*: Setup and preliminary results." *International Journal of Primatology* 31, no. 2 (2010): 159–180.

Berthaume, Michael A. "Tooth cusp sharpness as a dietary correlate in great apes." *American Journal of Physical Anthropology* 153, no. 2 (2014): 226–235.

Berthaume, Michael A., Lucas K. Delezene, and Kornelius Kupczik. "Dental topography and the diet of *Homo naledi*." *Journal of Human Evolution* 118 (2018): 14–26.

Berthaume, Michael [A.], Ian R. Grosse, Nirdesh D. Patel, David S. Strait, Sarah Wood, and Brian G. Richmond. "The effect of early hominin occlusal morphology on the fracturing of hard food items." *Anatomical Record: Advances in Integrative Anatomy and Evolutionary Biology* 293, no. 4 (2010): 594–606.

Bertram, John E. A., and Sharon M. Swartz. "The 'law of bone transformation': A case of crying Wolff?" *Biological Reviews* 66, no. 3 (1991): 245–273.

Betz, Oliver. "Ecomorphology: Integration of form, function, and ecology in the analysis of morphological structures." *Mitteilungen der Deutschen Gesellschaft für Allgemeine und Angewandte Entomologie* 15 (2006): 409–416.

Biewener, Andrew A. "Bone strength in small mammals and bipedal birds: Do safety factors change with body size?" *Journal of Experimental Biology* 98, no. 1 (1982): 289–301.

Biewener, Andrew A. "Safety factors in bone strength." *Calcified Tissue International* 53, no. 1 (1993): S68–S74.

Blob, Richard W., and Andrew A. Biewener. "*In vivo* locomotor strain in the hindlimb bones of *Alligator mississippiensis* and *Iguana iguana*: Implications for the evolution of limb bone safety factor and non-sprawling limb posture." *Journal of Experimental Biology* 202, no. 9 (1999): 1023–1046.

Blob, Richard W., Nora R. Espinoza, Michael T. Butcher, Andrew H. Lee, Angela R. D'Amico, Faraz Baig, and K. Megan Sheffield. "Diversity of limb-bone safety factors for locomotion in terrestrial vertebrates: Evolution and mixed chains." *Integrative and Comparative Biology* 54, no. 6 (2014): 1058–1071.

Blumenschine, Robert J. "Carcass consumption sequences and the archaeological distinction of scavenging and hunting." *Journal of Human Evolution* 15, no. 8 (1986): 639–659.

Bock, Walter J. "The role of adaptive mechanisms in the origin of higher levels of organization." *Systematic Zoology* 14, no. 4 (1965): 272–287.

Bock, Walter J. "The definition and recognition of biological adaptation." *American Zoologist* 20, no. 1 (1980): 217–227.

Bock, Walter J., and Gerd von Wahlert. "Adaptation and the form-function complex." *Evolution* 19, no. 3 (1965): 269–299.

Bookstein, Fred L. "'Like fixing an airplane in flight': On paleoanthropology as an evolutionary discipline; or, Paleoanthropology for what?" In *Rethinking Human Evolution*, edited by Jeffrey H. Schwartz, 181–202. Vienna Series in Theoretical Biology No. 21. Cambridge, MA, MIT Press, 2018.

Bouvier, Marianne. "Biomechanical scaling of mandibular dimensions in New World monkeys." *International Journal of Primatology* 7, no. 6 (1986): 551–567.

Bouvier, Marianne, and William L. Hylander. "Effect of bone strain on cortical bone structure in macaques (*Macaca mulatta*)." *Journal of Morphology* 167, no. 1 (1981): 1–12.

Bouvier, Marianne, and William L. Hylander. "The mechanical or metabolic function of secondary osteonal bone in the monkey *Macaca fascicularis*." *Archives of Oral Biology* 41, no. 10 (1996): 941–950.

Boyer, Doug M. "Relief index of second mandibular molars is a correlate of diet among prosimian primates and other euarchontan mammals." *Journal of Human Evolution* 55, no. 6 (2008): 1118–1137.

Bradfield, Justin. "Identifying animal taxa used to manufacture bone tools during the Middle Stone Age at Sibudu, South Africa: Results of a CT-rendered histological analysis." *PLoS One* 13, no. 11 (2018): e0208319.

Brain, Charles Kimberlin. *The Hunters or the Hunted? An Introduction to African Cave Taphonomy.* Chicago: University of Chicago Press, 1981.

Bramble, Dennis M., and Daniel E. Lieberman. "Endurance running and the evolution of *Homo*." *Nature* 432, no. 7015 (2004): 345–352.

Bribiescas, Richard G. "On the evolution, life history, and proximate mechanisms of human male reproductive senescence." *Evolutionary Anthropology: Issues, News, and Reviews* 15, no. 4 (2006): 132–141.

Bridges, Patricia S. "Changes in activities with the shift to agriculture in the southeastern United States." *Current Anthropology* 30, no. 3 (1989): 385–394.

Bridges, Patricia S. "Structural changes in the femur with the transition to agriculture on the Georgia coast." In *What Mean These Bones? Studies in Southeastern Bioarchaeology*, edited by Mary Lucas Powell, Patricia S. Bridges, and Ann Marie Wagner Mires, Ch. 7. Tuscaloosa: University of Alabama Press, 1991.

Bromage, Timothy G. "Ontogeny of the early hominid face." *Journal of Human Evolution* 18, no. 8 (1989): 751–773.

Brown, Alejandro D., and Gabriel E. Zunino. "Dietary variability in *Cebus apella* in extreme habitats: Evidence for adaptability." *Folia Primatologica* 54, no. 3–4 (1990): 187–195.

Brunet, Michel, Franck Guy, David Pilbeam, Hassane Taisso Mackaye, Andossa Likius, Djimdoumalbaye Ahounta, Alain Beauvilain, et al. "A new hominid from the Upper Miocene of Chad, Central Africa." *Nature* 418, no. 6894 (2002): 145–151

Bunn, Jonathan M., Doug M. Boyer, Yaron Lipman, Elizabeth M. St. Clair, Jukka Jernvall, and Ingrid Daubechies. "Comparing Dirichlet normal surface energy of tooth crowns, a new technique of molar shape quantification for dietary inference, with previous methods in isolation and in combination." *American Journal of Physical Anthropology* 145, no. 2 (2011): 247–261.

Burgman, Jenny H. E., Jennifer Leichliter, Nico L. Avenant, and Peter S. Ungar. "Dental microwear of sympatric rodent species sampled across habitats in southern Africa: Implications for environmental influence." *Integrative Zoology* 11, no. 2 (2016): 111–127.

Burr, David B. and Matthew R. Allen. *Basic and Applied Bone Biology*. New York: Academic Press, 2014.

Burr, David B., Alexander G. Robling, and Charles H. Turner. "Effects of biomechanical stress on bones in animals." *Bone* 30, no. 5 (2002): 781–786.

Butler, Percy M. "The evolution of tooth shape and tooth function in primates." In *Development, Function and Evolution of Teeth*, edited by Mark F. Teaford, Moya Meredith Smith, and Mark W. J. Ferguson, 201–211. Cambridge: Cambridge University Press, 2009.

Cachel, Susan M. "A new view of speciation in *Australopithecus*." In *Paleoanthropology: Morphology and Paleoecology*, edited by Russell H. Tuttle, 183–201. The Hague: Mouton, 1975.

Cachel, S[usan M.]. "Growth and allometry in primate masticatory muscles." *Archives of Oral Biology* 29, no. 4 (1984): 287–293.

Cardini, Andrea, and Sarah Elton. "Variation in guenon skulls (I): Species divergence, ecological and genetic differences." *Journal of Human Evolution* 54, no. 5 (2008): 615–637.

Cardini, Andrea, and Sarah Elton. "Geographical and taxonomic influences on cranial variation in red colobus monkeys (Primates, Colobinae): Introducing a new approach to 'morph' monkeys." *Global Ecology and Biogeography* 18, no. 2 (2009): 248–263.

Cardini, Andrea, Paul O'Higgins, and F. James Rohlf. "Seeing distinct groups where there are none: Spurious patterns from between-group PCA." *Evolutionary Biology* 46, no. 4 (2019): 303–316.

Careau, Vincent, and Theodore Garland Jr. "Performance, personality, and energetics: Correlation, causation, and mechanism." *Physiological and Biochemical Zoology* 85, no. 6 (2012): 543–571.

Carpenter, R. Dana, and Dennis R. Carter. "The mechanobiological effects of periosteal surface loads." *Biomechanics and Modeling in Mechanobiology* 7, no. 3 (2008): 227–242.

Carrier, David R., and Michael H. Morgan. "Protective buttressing of the hominin face." *Biological Reviews* 90, no. 1 (2015): 330–346.

Carter, Dennis R., and Gary S. Beaupré. *Skeletal Function and Form: Mechanobiology of Skeletal Development, Aging, and Regeneration*. Cambridge: Cambridge University Press, 2007.

Carter, D[ennis] R., D. P. Fyhrie, and R. T. Whalen. "Trabecular bone density and loading history: Regulation of connective tissue biology by mechanical energy." *Journal of Biomechanics* 20, no. 8 (1987): 785–794.

Cartmill, Matt. "Pads and claws in arboreal locomotion." In *Primate Locomotion*, edited by Farish A. Jenkins Jr., 45–83. New York: Academic Press, 1974.

Cartmill, Matt. "Human uniqueness and theoretical content in paleoanthropology." *International Journal of Primatology* 11, no. 3 (1990): 173–192.

Caspari, Rachel, and Milford H. Wolpoff. "The Dubois syndrome." *History and Philosophy of the Life Sciences* 34, no. 1–2 (2012): 33–42.

Cerling, Thure E., Fredrick Kyalo Manthi, Emma N. Mbua, Louise N. Leakey, Meave G. Leakey, Richard E. Leakey, Francis H. Brown, et al. "Stable isotope-based diet reconstructions of Turkana Basin hominins." *Proceedings of the National Academy of Sciences* 110, no. 26 (2013): 10501–10506.

Cerling, Thure E., Emma Mbua, Francis M. Kirera, Fredrick Kyalo Manthi, Frederick E. Grine, Meave G. Leakey, Matt Sponheimer, and Kevin T. Uno. "Diet of *Paranthropus boisei* in the early Pleistocene of East Africa." *Proceedings of the National Academy of Sciences* 108, no. 23 (2011): 9337–9341.

Chamberlain, Jack G., and Walter C. Hartwig. "Thomas Kuhn and paleoanthropology." *Evolutionary Anthropology: Issues, News, and Reviews* 8, no. 2 (1999): 42–44.

Charlesworth, Brian, Russell Lande, and Montgomery Slatkin. "A neo-Darwinian commentary on macroevolution." *Evolution* 36, no. 3 (1982): 474–498.

Cheverud, James M. "Phenotypic, genetic, and environmental morphological integration in the cranium." *Evolution* 36, no. 3 (1982): 499–516.

Cheverud, James M. "Quantitative genetics and developmental constraints on evolution by selection." *Journal of Theoretical Biology* 110, no. 2 (1984): 155–171.

Cheverud, James M. "Developmental integration and the evolution of pleiotropy." *American Zoologist* 36, no. 1 (1996): 44–50.

Chiu, Chi-Hua, and Mark W. Hamrick. "Evolution and development of the primate limb skeleton." *Evolutionary Anthropology: Issues, News, and Reviews* 11, no. 3 (2002): 94–107.

Churchill, Steven E., Osbjorn M. Pearson, Frederick E. Grine, Erik Trinkaus, and Trenton W. Holliday. "Morphological affinities of the proximal ulna from Klasies River main site: Archaic or modern?" *Journal of Human Evolution* 31, no. 3 (1996): 213–237.

Ciochon, R[ussell] L., R. A. Nisbett, and R. S. Corruccini. "Dietary consistency and craniofacial development related to masticatory function in minipigs." *Journal of Craniofacial Genetics and Developmental Biology* 17, no. 2 (1997): 96–102.

Ciochon, Russell L., Dolores R. Piperno, and Robert G. Thompson. "Opal phytoliths found on the teeth of the extinct ape *Gigantopithecus blacki*: Implications for paleodietary studies." *Proceedings of the National Academy of Sciences* 87, no. 20 (1990): 8120–8124.

Clark, W[ilfred] E. Le Gros. "Hominid characters of the australopithecine dentition." *Journal of the Royal Anthropological Institute of Great Britain and Ireland* 80, no. 1–2 (1950): 37–54.

Clark, Wilfrid E. Le Gros. *The Fossil Evidence for Human Evolution*. Chicago: University of Chicago Press, 1955.

Clark, Wilfred E. Le Gros. *The Antecedents of Man*. New York: Harper & Row, 1959.

Cobley, Matthew J., Emily J. Rayfield, and Paul M. Barrett. "Inter-vertebral flexibility of the ostrich neck: Implications for estimating sauropod neck flexibility." *PLoS One* 8, no. 8 (2013): e72187.

Codron, Daryl, Julia A. Lee-Thorp, Matt Sponheimer, Darryl de Ruiter, and Jacqueline Codron. "Inter- and intrahabitat dietary variability of chacma baboons (*Papio ursinus*) in South African savannas based on fecal δ13C, δ15N, and %N." *American Journal of Physical Anthropology* 129, no. 2 (2006): 204–214.

Collard, Mark, and Bernard Wood. "Hominin homoiology: An assessment of the impact of phenotypic plasticity on phylogenetic analyses of humans and their fossil relatives." *Journal of Human Evolution* 52, no. 5 (2007): 573–584.

Congdon, Kimberly A. "Interspecific and ontogenetic variation in proximal pedal phalangeal curvature of great apes (*Gorilla gorilla*, *Pan troglodytes*, and *Pongo pygmaeus*)." *International Journal of Primatology* 33, no. 2 (2012): 418–427.

Conklin-Brittain, Nancy Lou, Richard W. Wrangham, and Kevin D. Hunt. "Dietary response of chimpanzees and cercopithecines to seasonal variation in fruit abundance, II: Macronutrients." *International Journal of Primatology* 19, no. 6 (1998): 971–998.

Constantino, Paul J., James J.-W. Lee, Yvonne Gerbig, Adam Hartstone-Rose, Mauricio Talebi, Brian R. Lawn, and Peter W. Lucas. "The role of tooth enamel mechanical properties in primate dietary adaptation." *American Journal of Physical Anthropology* 148, no. 2 (2012): 171–177.

Cooper, W. James, and Scott J. Steppan. "Developmental constraint on the evolution of marsupial forelimb morphology." *Australian Journal of Zoology* 58, no. 1 (2010): 1–15.

Copeland, Sandi R., Matt Sponheimer, Darryl J. de Ruiter, Julia A. Lee-Thorp, Daryl Codron, Petrus J. le Roux, Vaughan Grimes, and Michael P. Richards. "Strontium isotope evidence for landscape use by early hominins." *Nature* 474, no. 7349 (2011): 76–78.

Coquerelle, Michael, Juan Carlos Prados-Frutos, Rosa Rojo, Philipp Mitteroecker, Markus Bastir, and David Frayer. "Short faces, big tongues: Developmental origin of the human chin." *PLoS One* 8, no. 11 (2013): e81287.

Corruccini, Robert S., and Henry M. McHenry. "Knuckle-walking hominid ancestors." *Journal of Human Evolution* 40, no. 6 (2001): 507–511.

Cowin, S. C., R. T. Hart, J. R. Balser, and D. H. Kohn. "Functional adaptation in long bones: Establishing *in vivo* values for surface remodeling rate coefficients." *Journal of Biomechanics* 18, no. 9 (1985): 665–684.

Crispo, Erika. "The Baldwin effect and genetic assimilation: Revisiting two mechanisms of evolutionary change mediated by phenotypic plasticity." *Evolution: International Journal of Organic Evolution* 61, no. 11 (2007): 2469–2479.

Crompton, Robin Huw, Juliet McClymont, Susannah Thorpe, William Sellers, Jason Heaton, Travis Rayne Pickering, Todd Pataky, et al. "Functional anatomy, biomechanical performance capabilities and potential niche of StW 573: An *Australopithecus* skeleton (circa 3.67 Ma) from Sterkfontein Member 2, and its significance for the last common ancestor of the African apes and for hominin origins." *BioRxiv* (2018): 481556.

Crompton, Robin Huw, Li Yu, Wang Weijie, Michael Günther, and Russell Savage. "The mechanical effectiveness of erect and 'bent-hip, bent-knee' bipedal walking in *Australopithecus afarensis*." *Journal of Human Evolution* 35, no. 1 (1998): 55–74.

Cubo, Jorge, Pierre Legendre, Armand De Ricqlès, Laëtitia Montes, Emmanual De Margerie, Jacques Castanet, and Yves Desdevises. "Phylogenetic, functional, and structural components of variation in bone growth rate of amniotes." *Evolution & Development* 10, no. 2 (2008): 217–227.

Cullen, D. M., R. T. Smith, and M. P. Akhter. "Bone-loading response varies with strain magnitude and cycle number." *Journal of Applied Physiology* 91, no. 5 (2001): 1971–1976.

Curio, Eberhard. "Towards a methodology of teleonomy." *Experientia* 29, no. 9 (1973): 1045–1058.

Currey, John D. "Mechanical properties of bone tissues with greatly differing functions." *Journal of Biomechanics* 12, no. 4 (1979): 313–319.

Daegling, David J. "Functional morphology of the human chin." *Evolutionary Anthropology: Issues, News, and Reviews* 1, no. 5 (1993): 170–177.

Daegling, David J. "Growth in the mandibles of African apes." *Journal of Human Evolution* 30, no. 4 (1996): 315–341.

Daegling, David J. "Relationship of strain magnitude to morphological variation in the primate skull." *American Journal of Physical Anthropology* 124, no. 4 (2004): 346–352.

Daegling, David J. "The human mandible and the origins of speech." *Journal of Anthropology* 2012 (2012): 1–14.

Daegling, David J., Henna D. Bhramdat, and Viviana Toro-Ibacache. "Efficacy of shear strain gradients as an osteogenic stimulus." *Journal of Theoretical Biology*, 524 (2021): 110730.

Daegling, David J., Kristian J. Carlson, Paul Tafforeau, Darryl J. de Ruiter, and Lee R. Berger. "Comparative biomechanics of *Australopithecus sediba* mandibles." *Journal of Human Evolution* 100 (2016): 73–86.

Daegling, David J., and Frederick E. Grine. "Compact bone distribution and biomechanics of early hominid mandibles." *American Journal of Physical Anthropology* 86, no. 3 (1991): 321–339.

Daegling, David J., and Frederick E. Grine. "Mandibular biomechanics and the paleontological evidence for the evolution of human diet." In *Evolution of the Human Diet: The Known, the Unknown, and the Unknowable*, edited by Peter S. Ungar, 77–105. New York: Oxford University Press, 2006.

Daegling, David J., and Frederick E. Grine. "Feeding behavior and diet in *Paranthropus boisei*: The limits of functional inference from the mandible." In *Human Paleontology and Prehistory*, edited by Assaf Marom and Erella Hovers, 109–125. Cham, Switzerland: Springer, 2017.

Daegling, David J., Stefan Judex, Engin Ozcivici, Matthew J. Ravosa, Andrea B. Taylor, Frederick E. Grine, Mark F. Teaford, and Peter S. Ungar. "Viewpoints: Feeding mechanics, diet, and dietary adaptations in early hominins." *American Journal of Physical Anthropology* 151, no. 3 (2013): 356–371.

Daegling, David J., and W. Scott McGraw. "Feeding, diet, and jaw form in West African *Colobus* and *Procolobus*." *International Journal of Primatology* 22, no. 6 (2001): 1033–1055.

Daegling, David J., Biren A. Patel, and William L. Jungers. "Geometric properties and comparative biomechanics of *Homo floresiensis* mandibles." *Journal of Human Evolution* 68 (2014): 36–46.

Dainton, Mike. "Did our ancestors knuckle-walk?" *Nature* 410, no. 6826 (2001): 324–325.

D'Amelio, Patrizia, Anastasia Grimaldi, Stefania Di Bella, Stefano Z. M. Brianza, Maria Angela Cristofaro, Cristina Tamone, Giuliana Giribaldi, Daniela Ulliers, Gian P. Pescarmona, and Giancarlo Isaia. "Estrogen deficiency increases osteoclastogenesis up-regulating T cells activity: A key mechanism in osteoporosis." *Bone* 43, no. 1 (2008): 92–100.

Dart, Raymond A[rthur]. "*Australopithecus africanus*: The man-ape of South Africa." *Nature* 115, no. 2884 (1925): 195–199.

Dart, Raymond A[rthur]. "The Makapansgat proto-human *Australopithecus prometheus*." *American Journal of Physical Anthropology* 6, no. 3 (1948): 259–283.

Dart, Raymond [Arthur]. "The osteodontokeratic culture of *Australopithecus prometheus*: Discussion." *Transvaal Museum Memoirs* 10, no. 1 (1957): 87–101.

Dart, Raymond Arthur, and Dennis Craig. *Adventures with the Missing Link*. London: H. Hamilton, 1959.

Day, Michael H. *Guide to Fossil Man*. Chicago: University of Chicago Press, 1977.

DeGusta, David, W. Henry Gilbert, and Scott P. Turner. "Hypoglossal canal size and hominid speech." *Proceedings of the National Academy of Sciences* 96, no. 4 (1999): 1800–1804.

Demes, B[rigitte]. "Biomechanics of the primate skull base." *Advances in Anatomy, Embryology, and Cell Biology* 94 (1985): 1–59.

Demes, Brigitte, and Norman Creel. "Bite force, diet, and cranial morphology of fossil hominids." *Journal of Human Evolution* 17, no. 7 (1988): 657–670.

Demes, Brigitte, William L. Jungers, and Christopher Walker. "Cortical bone distribution in the femoral neck of strepsirhine primates." *Journal of Human Evolution* 39, no. 4 (2000): 367–379.

Demes, B[rigitte], H. Preuschoft, and J. E. A. Wolff. "Stress-strength relationships in the mandibles of hominoids." In *Food Acquisition and Processing in Primates*, edited by David Chivers, Bernard Wood, and Andrew Bilsborough, 369–390. Boston: Springer, 1984.

Demes, Brigitte, Yi-Xian Qin, Jack T. Stern Jr., Susan G. Larson, and Clinton T. Rubin. "Patterns of strain in the macaque tibia during functional activity." *American Journal of Physical Anthropology* 116, no. 4 (2001): 257–265.

Demes, Brigitte, Jack T. Stern Jr., Michael R. Hausman, Susan G. Larson, Kenneth J. McLeod, and Clinton T. Rubin. "Patterns of strain in the macaque ulna during functional activity." *American Journal of Physical Anthropology* 106, no. 1 (1998): 87–100.

de Ruiter, Darryl J., Thomas J. DeWitt, Keely B. Carlson, Juliet K. Brophy, Lauren Schroeder, Rebecca R. Ackermann, Steven E. Churchill, and Lee R. Berger. "Mandibular remains support taxonomic validity of *Australopithecus sediba*." *Science* 340, no. 6129 (2013): 164.

DeVore, Irven, and Sherwood L. Washburn. "Baboon ecology and human evolution." In *African Ecology and Human Evolution*, edited by F. Clark Howell and François Bourlière, 335–367. New York: Wenner-Gren Foundation, 1963.

Disotell, Todd R. "Generic level relationships of the Papionini (Cercopithecoidea)." *American Journal of Physical Anthropology* 94, no. 1 (1994): 47–57.

Doran, Diane M. "The ontogeny of chimpanzee and pygmy chimpanzee locomotor behavior: A case study of paedomorphism and its behavioral correlates." *Journal of Human Evolution* 23, no. 2 (1992): 139–157.

Drapeau, Michelle S. M. "Functional anatomy of the olecranon process in hominoids and Plio-Pleistocene hominins." *American Journal of Physical Anthropology* 124, no. 4 (2004): 297–314.

Du Brul, E. Lloyd. "Early hominid feeding mechanisms." *American Journal of Physical Anthropology* 47, no. 2 (1977): 305–320.

Du Brul, E. Lloyd, and Harry Sicher. *The Adaptive Chin.* Springfield, IL: Charles C. Thomas, 1954.

Dullemeijer, Pieter. "Functional morphology and evolutionary biology." *Acta Biotheoretica* 29, no. 3–4 (1980): 151–250.

Dullemeijer, Pieter. "Diversity of functional morphological explanation." In *Architecture in Living Structure*, edited by Gart A. Zweers and Pieter Dullemeijer, 5–17. Dordrecht, Netherlands: Springer, 1985.

Duncan, Alexander S., John Kappelman, and Liza J. Shapiro. "Metatarsophalangeal joint function and positional behavior in *Australopithecus afarensis*." *American Journal of Physical Anthropology* 93, no. 1 (1994): 67–81.

Dwyer, Peter D. "Functionalism and structuralism: Two programs for evolutionary biologists." *American Naturalist* 124, no. 5 (1984): 745–750.

Eble, Gunther J. "Theoretical morphology: State of the art [review of *Theoretical Morphology: The Concept and Its Applications*, edited by George R. McGhee Jr.]." *Paleobiology* 26, no. 3 (2000): 520–528.

Eisenberg, John Frederick. *The Mammalian Radiations: An Analysis of Trends in Evolution, Adaptation, and Behavior.* Chicago: University of Chicago Press, 1981.

Eklöv, Peter, and Richard Svanbäck. "Predation risk influences adaptive morphological variation in fish populations." *American Naturalist* 167, no. 3 (2006): 440–452.

Endo, Banri. "Stress analysis on the facial skeleton of gorilla by means of the wire strain gauge method." *Primates* 14, no. 1 (1973): 37–45.

Eng, Carolyn M., Daniel E. Lieberman, Katherine D. Zink, and Michael A. Peters. "Bite force and occlusal stress production in hominin evolution." *American Journal of Physical Anthropology* 151, no. 4 (2013): 544–557.

Erickson, Gregory M., Joseph Catanese III, and Tony M. Keaveny. "Evolution of the biomechanical material properties of the femur." *Anatomical Record: Advances in Integrative Anatomy and Evolutionary Biology* 268, no. 2 (2002): 115–124.

Fannin, Luke D., and W. Scott McGraw. "Circumorbital rim variation in western red colobus (*Piliocolobus badius badius*) and its potential role as a sexually selected trait." *American Journal of Physical Anthropology* 165 (2018): 81–82.

Fannin, Luke D., J. Michael Plavcan, David J. Daegling, and W. Scott McGraw. "Oral processing, sexual selection, and size variation in the circumorbital region of *Colobus* and *Piliocolobus*." *American Journal of Physical Anthropology*, 2021 (April): 1–18, https://doi.org/10.1002/ajpa.24280/.

Farris, Dominic James, Jonathon Birch, and Luke Kelly. "Foot stiffening during the push-off phase of human walking is linked to active muscle contraction, and not the windlass mechanism." *Journal of the Royal Society Interface* 17, no. 168 (2020): article ID 20200208.

Feibel, Craig S., John M. Harris, and Francis H. Brown. "Palaeoenvironmental context for the late Neogene of the Turkana Basin." *Koobi Fora Research Project*, vol. 3, edited by J. M. Harris, 321–346. New York: Oxford University Press, 1991.

Fisher, Daniel C. "Evolutionary morphology: Beyond the analogous, the anecdotal, and the ad hoc." *Paleobiology* 11, no. 1 (1985): 120–138.

Fleagle, John G. "Locomotor behavior and skeletal anatomy of sympatric Malaysian leaf-monkeys (*Presbytis obscura* and *Presbytis melalophos*)." *Yearbook of Physical Anthropology* 20 (1977): 440–453.

Fleagle, John G. "Locomotor adaptations of Oligocene and Miocene hominoids and their phyletic implications." In *New Interpretations of Ape and Human Ancestry*, edited by Russell L. Ciochon and Robert S. Corruccini, 301–324. Boston: Springer, 1983.

Fleagle, John G. "Size and adaptation in primates." In *Size and Scaling in Primate Biology*, edited by William L. Jungers, 1–19. Boston: Springer, 1985.

Fleagle, John G. *Primate Adaptation and Evolution.* New York: Academic Press, 2013.

Fleagle, John G., and W. Scott McGraw. "Skeletal and dental morphology of African papionins: Unmasking a cryptic clade." *Journal of Human Evolution* 42, no. 3 (2002): 267–292.

Fleagle, John G., Jack T. Stern, William L. Jungers, Randall L. Susman, Andrea K. Vangor, and James P. Wells. "Climbing: A biomechanical link with brachiation and with bipedalism." In *Vertebrate Locomotion*, edited by Michael H. Day, 359–375. Symposia of the Zoological Society of London No. 48. London: Academic Press, 1981.

Foley, Cydne D., Arthur O. Quanbury, and T. Steinke. "Kinematics of normal child locomotion—a statistical study based on TV data." *Journal of Biomechanics* 12, no. 1 (1979): 1–8.

Forwood, Mark R., and David B. Burr. "Physical activity and bone mass: Exercises in futility?" *Bone and Mineral* 21, no. 2 (1993): 89–112.

Forwood, Mark R., and Charles H. Turner. "The response of rat tibiae to incremental bouts of mechanical loading: A quantum concept for bone formation." *Bone* 15, no. 6 (1994): 603–609.

Frost, Harold M. "Bone 'mass' and the 'mechanostat': A proposal." *Anatomical Record: Advances in Integrative Anatomy and Evolutionary Biology* 219, no. 1 (1987): 1–9.

Frost, Harold M. "Bone's mechanostat: A 2003 update." *Anatomical Record Part A: Discoveries in Molecular, Cellular, and Evolutionary Biology* 275, no. 2 (2003): 1081–1101.

Frost, Stephen R., Leslie F. Marcus, Fred L. Bookstein, David P. Reddy, and Eric Delson. "Cranial allometry, phylogeography, and systematics of large-bodied papionins (primates: Cercopithecinae) inferred from geometric morphometric analysis of landmark data." *Anatomical Record Part A: Discoveries in Molecular, Cellular, and Evolutionary Biology* 275, no. 2 (2003): 1048–1072.

Fukumoto, Seiji, and T. John Martin. "Bone as an endocrine organ." *Trends in Endocrinology & Metabolism* 20, no. 5 (2009): 230–236.

Gans, Carl. "Differences and similarities: Comparative methods in mastication." *American Zoologist* 25, no. 2 (1985): 291–302.

Gans, Carl. "Adaptation and the form-function relation." *American Zoologist* 28, no. 2 (1988): 681–697.

Garland, Theodore, Jr., and Scott A. Kelly. "Phenotypic plasticity and experimental evolution." *Journal of Experimental Biology* 209, no. 12 (2006): 2344–2361.

Garland, Theodore, Jr., Meng Zhao, and Wendy Saltzman. "Hormones and the evolution of complex traits: Insights from artificial selection on behavior." *Integrative and Comparative Biology* 56, no. 2 (2016): 207–224.

Garman, Russell, Clinton Rubin, and Stefan Judex. "Small oscillatory accelerations, independent of matrix deformations, increase osteoblast activity and enhance bone morphology." *PLoS One* 2, no. 7 (2007): e653.

Gatesy, S. M., and A. A. Biewener. "Bipedal locomotion: Effects of speed, size and limb posture in birds and humans." *Journal of Zoology* 224, no. 1 (1991): 127–147.

Gebo, Daniel L. "Climbing, brachiation, and terrestrial quadrupedalism: Historical precursors of hominid bipedalism." *American Journal of Physical Anthropology* 101, no. 1 (1996): 55–92.

Gebo, Daniel L. *Primate Comparative Anatomy.* Baltimore: Johns Hopkins University Press, 2014.

German, Rebecca Z., D. W. Hertweck, Joyce E. Sirianni, and Daris R. Swindler. "Heterochrony and sexual dimorphism in the pigtailed macaque (*Macaca nemestrina*)." *American Journal of Physical Anthropology* 93, no. 3 (1994): 373–380.

Ghalambor, Cameron K., John K. McKay, Scott P. Carroll, and David N. Reznick. "Adaptive versus non-adaptive phenotypic plasticity and the potential for contemporary adaptation in new environments." *Functional Ecology* 21, no. 3 (2007): 394–407.

Gould, Stephen Jay. *Ontogeny and Phylogeny.* Cambridge, MA: Harvard University Press, 1977.

Gould, Stephen Jay, and Niles Eldredge. "Punctuated equilibria: The tempo and mode of evolution reconsidered." *Paleobiology* 3, no. 2 (1977): 115–151.

Gould, Stephen Jay, and Richard C. Lewontin. "The spandrels of San Marco and the Panglossian paradigm: A critique of the adaptationist programme." *Proceedings of the Royal Society of London, Series B, Biological Sciences* 205, no. 1161 (1979): 581–598.

Gould, Stephen Jay, and Elisabeth S. Vrba. "Exaptation—a missing term in the science of form." *Paleobiology* 8, no. 1 (1982): 4–15.

Grabowski, Mark, Kevin G. Hatala, William L. Jungers, and Brian G. Richmond. "Body mass estimates of hominin fossils and the evolution of human body size." *Journal of Human Evolution* 85 (2015): 75–93.

Granatosky, Michael C., Eric J. McElroy, Myra F. Laird, Jose Iriarte-Diaz, Stephen M. Reilly, Andrea B. Taylor, and Callum F. Ross. "Joint angular excursions during cyclical behaviors differ between tetrapod feeding and locomotor systems." *Journal of Experimental Biology* 222, no. 9 (2019): jeb200451.

Greaves, W. S. "The jaw lever system in ungulates: A new model." *Journal of Zoology* 184, no. 2 (1978): 271–285.

Green, David J., and Zeresenay Alemseged. "*Australopithecus afarensis* scapular ontogeny, function, and the role of climbing in human evolution." *Science* 338, no. 6106 (2012): 514–517.

Green, Richard E., Johannes Krause, Adrian W. Briggs, Tomislav Maricic, Udo Stenzel, Martin Kircher, Nick Patterson, et al. "A draft sequence of the Neandertal genome." *Science* 328, no. 5979 (2010): 710–722.

Greene, Harry W. "Homology and behavioral repertoires." In *Homology: The Hierarchial Basis of Comparative Biology*, edited by Brian K. Hall, 369–390. San Diego: Academic Press, 1994.

Grine, Frederick E. "Trophic differences between 'gracile' and 'robust' australopithecines: A scanning electron microscope analysis of occlusal events." *South African Journal of Science* 77, no. 5 (1981): 203–230.

Grine, Frederick E. "Dental evidence for dietary differences in *Australopithecus* and *Paranthropus*: A quantitative analysis of permanent molar microwear." *Journal of Human Evolution* 15, no. 8 (1986): 783–822.

Grine, Frederick E., and David J. Daegling. "Functional morphology, biomechanics and the retrodiction of early hominin diets." *Comptes Rendus Palevol* 16, no. 5–6 (2017): 613–631.

Grine, Frederick E., Stefan Judex, David J. Daegling, Engin Ozcivici, Peter S. Ungar, Mark F. Teaford, Matt Sponheimer, et al. "Craniofacial biomechanics and functional and dietary inferences in hominin paleontology." *Journal of Human Evolution* 58, no. 4 (2010): 293–308.

Grine, Frederick E., and Richard F. Kay. "Early hominid diets from quantitative image analysis of dental microwear." *Nature* 333, no. 6175 (1988): 765–768.

Gröning, Flora, Michael Fagan, and Paul O'Higgins. "Modeling the human mandible under masticatory loads: Which input variables are important?" *Anatomical Record: Advances in Integrative Anatomy and Evolutionary Biology* 295, no. 5 (2012): 853–863.

Gross, Ted S., Jonathan L. Edwards, Kenneth J. McLeod, and Clinton T. Rubin. "Strain gradients correlate with sites of periosteal bone formation." *Journal of Bone and Mineral Research* 12, no. 6 (1997): 982–988.

Grove, Matt. "Change and variability in Plio-Pleistocene climates: Modelling the hominin response." *Journal of Archaeological Science* 38, no. 11 (2011a): 3038–3047.

Grove, Matt. "Speciation, diversity, and Mode 1 technologies: The impact of variability selection." *Journal of Human Evolution* 61, no. 3 (2011b): 306–319.

Grove, Matt. "Evolution and dispersal under climatic instability: A simple evolutionary algorithm." *Adaptive Behavior* 22, no. 4 (2014): 235–254.

Groves, Colin P., and John R. Napier. "Dental dimensions and diet in australopithecines." In *Proceedings of the VIIIth International Congress of Anthropological and Ethnological Sciences, 1968, Tokyo and Kyoto*, vol. 3, 273–276. Tokyo: Science Council of Japan, 1969.

Gruss, Laura Tobias, and Daniel Schmitt. "The evolution of the human pelvis: Changing adaptations to bipedalism, obstetrics and thermoregulation." *Philosophical Transactions of the Royal Society B: Biological Sciences* 370, no. 1663 (2015): article ID 20140063.

Gunness, M., and J. M. Hock. "Anabolic effect of parathyroid hormone on cancellous and cortical bone histology." *Bone* 14, no. 3 (1993): 277–281.

Gunter, Katherine B., Hawley C. Almstedt, and Kathleen F. Janz. "Physical activity in childhood may be the key to optimizing lifespan skeletal health." *Exercise and Sport Sciences Reviews* 40, no. 1 (2012): 13–21.

Haile-Selassie, Yohannes, Bruce M. Latimer, Mulugeta Alene, Alan L. Deino, Luis Gibert, Stephanie M. Melillo, Beverly Z. Saylor, Gary R. Scott, and C. Owen Lovejoy. "An early *Australopithecus afarensis* postcranium from Woranso-Mille, Ethiopia." *Proceedings of the National Academy of Sciences* 107, no. 27 (2010): 12121–12126.

Hall, Brian K. "Organic selection: Proximate environmental effects on the evolution of morphology and behaviour." *Biology and Philosophy* 16, no. 2 (2001): 215–237.

Hall, Brian K. "Palaeontology and evolutionary developmental biology: A science of the nineteenth and twenty-first centuries." *Palaeontology* 45, no. 4 (2002): 647–669.

Hall, Brian K. "Homoplasy and homology: Dichotomy or continuum?" *Journal of Human Evolution* 52, no. 5 (2007): 473–479.

Hall, Roberta L., and Henry S. Sharp. *Wolf and Man: Evolution in Parallel*. New York: Academic Press, 1978.

Hallgrímsson, Benedikt, and Brian K. Hall. "Variation and variability: Central concepts in biology." In *Variation: A Central Concept in Biology*, 1–7. New York: Elsevier/Academic Press, 2011.

Hallgrímsson, Benedikt, Katherine Willmore, and Brian K. Hall. "Canalization, developmental stability, and morphological integration in primate limbs." *American Journal of Physical Anthropology* 119, no. S35 (2002): 131–158.

Hamrick, Mark W. "The developmental origins of mosaic evolution in the primate limb skeleton." *Evolutionary Biology* 39, no. 4 (2012): 447–455.

Hamrick, Mark W., and Serge Livio Ferrari. "Leptin and the sympathetic connection of fat to bone." *Osteoporosis International* 19, no. 7 (2008): 905–912.

Hanna, Jandy B., Daniel Schmitt, and Timothy M. Griffin. "The energetic cost of climbing in primates." *Science* 320, no. 5878 (2008): 898–898.

Hardy, Alister. "Was man more aquatic in the past?" *New Scientist* 7, no. 5 (1960): 642–645.

Hartwig, Walter Carl. *The Primate Fossil Record*. Cambridge: Cambridge University Press, 2002.

Harvey, Paul H., and Timothy H. Clutton-Brock. "Life history variation in primates." *Evolution* 39, no. 3 (1985): 559–581.

Henke, Winfried. "Human biological evolution." In *Handbook of Evolution*, vol. 2, edited by Franz M. Wuketits and Christoph Antweiler, 117–222. Weinheim, Germany: Wiley-VCH, 2005.

Henry, Amanda G., Peter S. Ungar, Benjamin H. Passey, Matt Sponheimer, Lloyd Rossouw, Marion Bamford, Paul Sandberg, Darryl J. de Ruiter, and Lee Berger. "The diet of *Australopithecus sediba*."*Nature* 487, no. 7405 (2012): 90–93.

Hershkovitz, P. "The decorative chin." *Bulletin of the Field Museum of Natural History* 41 (1970): 6–10.

Hlusko, Leslea J. "Integrating the genotype and phenotype in hominid paleontology." *Proceedings of the National Academy of Sciences* 101, no. 9 (2004): 2653–2657.

Hlusko, Leslea J., Gen Suwa, Reiko T. Kono, and Michael C. Mahaney. "Genetics and the evolution of primate enamel thickness: A baboon model." *American Journal of Physical Anthropology* 124, no. 3 (2004): 223–233.

Hodge, W. A., R. S. Fijan, K. L. Carlson, R. G. Burgess, W. H. Harris, and R. W. Mann. "Contact pressures in the human hip joint measured *in vivo*." *Proceedings of the National Academy of Sciences* 83, no. 9 (1986): 2879–2883.

Holliday, Trenton W., Steven E. Churchill, Kristian J. Carlson, Jeremy M. DeSilva, Peter Schmid, Christopher S. Walker, and Lee R. Berger. "*Australopithecus sediba*—body size and proportions of *Australopithecus sediba*." *PaleoAnthropology* 406 (2018): 422.

Hsieh, Yeou-Fang, Alexander G. Robling, Walter T. Ambrosius, David B. Burr, and Charles H. Turner. "Mechanical loading of diaphyseal bone *in vivo*: The strain threshold for an osteogenic response varies with location." *Journal of Bone and Mineral Research* 16, no. 12 (2001): 2291–2297.

Hua, Licheng, Jianbin Chen, and Peter S. Ungar. "Diet reduces the effect of exogenous grit on tooth microwear." *Biosurface and Biotribology* 6, no. 2 (2020): 48–52.

Huiskes, R. "If bone is the answer, then what is the question?" *Journal of Anatomy* 197, no. 2 (2000): 145–156.

Huiskes, R., and E. Y. S. Chao. "A survey of finite element analysis in orthopedic biomechanics: The first decade." *Journal of Biomechanics* 16, no. 6 (1983): 385–409.

Hunt, Kevin D. "The evolution of human bipedality: Ecology and functional morphology." *Journal of Human Evolution* 26, no. 3 (1994): 183–202.

Hunt, Kevin D. "Ecological morphology of *Australopithecus afarensis*: Traveling terrestrially, eating arboreally." In *Primate Locomotion: Recent Advances*, edited by Elizabeth Strasser, John G. Fleagle, Alfred L. Rosenberger, and Henry M. McHenry, 397–418. Boston: Springer, 1998.

Hutchinson, John R., and Stephen M. Gatesy. "Bipedalism." *Encyclopedia of Life Sciences* (April 2001): 1–6.

Hutson, Joel D., and Kelda N. Hutson. "Using the American alligator and a repeated-measures design to place constraints on *in vivo* shoulder joint range of motion in dinosaurs and other fossil archosaurs." *Journal of Experimental Biology* 216, no. 2 (2013): 275–284.

Huxley, Julian S. "Evolutionary processes and taxonomy with special reference to grades." *Uppsala Universitets Årsskrift* 6 (1958): 21–39.

Hylander, William L. "Incisor size and diet in anthropoids with special reference to Cercopithecidae." *Science* 189, no. 4208 (1975): 1095–1098.

Hylander, William L. "The functional significance of primate mandibular form." *Journal of Morphology* 160, no. 2 (1979): 223–239.

Hylander, William L. "Stress and strain in the mandibular symphysis of primates: A test of competing hypotheses." *American Journal of Physical Anthropology* 64, no. 1 (1984): 1–46.

Hylander, William L. "Mandibular function and biomechanical stress and scaling." *American Zoologist* 25, no. 2 (1985): 315–330.

Hylander, William L. "Implications of *in vivo* experiments for interpreting the functional significance of 'robust' australopithecine jaws." In *Evolutionary History of the "Robust" Australopithecines*, edited by Frederick E. Grine, 55–83. New York: Aldine de Gruyter, 1988.

Hylander, William L., and Kirk R. Johnson. "Strain gradients in the craniofacial region of primates." In *Biological Mechanisms of Tooth Movement and Craniofacial Adaptation*, edited by Zeev Davidovitch, 559–569. Columbus: Ohio State University, College of Dentistry, 1992.

Hylander, William L., and Kirk R. Johnson. "Functional morphology and *in vivo* bone strain patterns in the craniofacial region of primates: Beware of biomechanical stories about fossil bones." In *Reconstructing Behavior in the Primate Fossil Record*, edited by J. Michael Plavcan, Richard F. Kay, William L. Jungers, and Carel P. van Schaik, 43–72. Boston: Springer, 2002.

Ichim, Ionut, Jules Kieser, and Michael Swain. "Tongue contractions during speech may have led to the development of the bony geometry of the chin following the evolution of human language: A mechanobiological hypothesis for the development of the human chin." *Medical Hypotheses* 69, no. 1 (2007): 20–24.

Inouye, Sandra E., and Brian T. Shea. "What's your angle? Size correction and bar-glenoid orientation in 'Lucy' (AL 288-1)." *International Journal of Primatology* 18, no. 4 (1997): 629–650.

Jablonski, Nina G., and George Chaplin. "Origin of habitual terrestrial bipedalism in the ancestor of the Hominidae." *Journal of Human Evolution* 24, no. 4 (1993): 259–280.

Jablonski, Nina G., and George Chaplin. "The fossil record of gibbons." In *The Gibbons*, edited by Danielle Whittaker and Susan Lappan, 111–130. New York: Springer, 2009.

Johanson, Donald C., C. Owen Lovejoy, William H. Kimbel, Tim D. White, Steven C. Ward, Michael E. Bush, Bruce M. Latimer, and Yves Coppens. "Morphology of the Pliocene partial hominid skeleton (AL 288-1) from the Hadar Formation, Ethiopia." *American Journal of Physical Anthropology* 57, no. 4 (1982): 403–451.

Jolly, Clifford J. "The seed-eaters: A new model of hominid differentiation based on a baboon analogy." *Man* 5, no. 1 (1970): 5–26.

Judex, Stefan, and Kristian J. Carlson. "Is bone's response to mechanical signals dominated by gravitational loading?" *Medicine and Science in Sports and Exercise* 41, no. 11 (2009): 2037–2043.

Judex, Stefan, Russell Garman, Maria Squire, Leah-Rae Donahue, and Clinton Rubin. "Genetically based influences on the site-specific regulation of trabecular and cortical bone morphology." *Journal of Bone and Mineral Research* 19, no. 4 (2004): 600–606.

Judex, Stefan, Ted S. Gross, and Ronald F. Zernicke. "Strain gradients correlate with sites of exercise-induced bone-forming surfaces in the adult skeleton." *Journal of Bone and Mineral Research* 12, no. 10 (1997): 1737–1745.

Jungers, William L. "Lucy's limbs: Skeletal allometry and locomotion in *Australopithecus afarensis*." *Nature* 297, no. 5868 (1982): 676–678.

Jungers, William L. "Aspects of size and scaling in primate biology with special reference to the locomotor skeleton." *American Journal of Physical Anthropology* 27, no. S5 (1984): 73–97.

Jungers, William L. "New estimates of body size in australopithecines." In *Evolutionary History of the "Robust" Australopithecines*, edited by Frederick E. Grine, 115–125. New York: Aldine de Gruyter, 1988a.

Jungers, William L. "Relative joint size and hominoid locomotor adaptations with implications for the evolution of hominid bipedalism." *Journal of Human Evolution* 17, no. 1–2 (1988b): 247–265.

Jungers, William [L.], and Karen Baab. "The geometry of hobbits: *Homo floresiensis* and human evolution." *Significance* 6, no. 4 (2009): 159–164.

Jungers, William L., Amy A. Pokempner, Richard F. Kay, and Matt Cartmill. "Hypoglossal canal size in living hominoids and the evolution of human speech." *Human Biology* 75, no. 4 (2003): 473–484.

Jungers, William L., and Jack T. Stern Jr. "Body proportions, skeletal allometry and locomotion in the Hadar hominids: A reply to Wolpoff." *Journal of Human Evolution* 12, no. 7 (1983): 673–684.

Jurmain, Robert. "Skeletal evidence of trauma in African apes, with special reference to the Gombe chimpanzees." *Primates* 38, no. 1 (1997): 1–14.

Kahneman, Daniel. *Thinking, Fast and Slow*. New York: Farrar, Straus & Giroux, 2011.

Kannus, Pekka, Heidi Haapasalo, Marja Sankelo, Harri Sievanen, Matti Pasanen, Ari Heinonen, Pekka Oja, and Ilkka Vuori. "Effect of starting age of physical activity on bone mass in the dominant arm of tennis and squash players." *Annals of Internal Medicine* 123, no. 1 (1995): 27–31.

Kappeler, Peter M., and Claudia Fichtel. "Eco-evo-devo of the lemur syndrome: Did adaptive behavioral plasticity get canalized in a large primate radiation?" *Frontiers in Zoology* 12, no. S1 (2015): S15.

Kappelman, John. "Plio-Pleistocene environments of Bed I and Lower Bed II, Olduvai Gorge, Tanzania." *Palaeogeography, Palaeoclimatology, Palaeoecology* 48, no. 2–4 (1984): 171–196.

Kay, Richard F. "The functional adaptations of primate molar teeth." *American Journal of Physical Anthropology* 43, no. 2 (1975): 195–215.

Kay, Richard F. "The nut-crackers—a new theory of the adaptations of the Ramapithecinae." *American Journal of Physical Anthropology* 55, no. 2 (1981): 141–151.

Kay, Richard F., and Matt Cartmill. "Cranial morphology and adaptations of *Palaechthon nacimienti* and other Paromomyidae (Plesiadapoidea, ? primates), with a description of a new genus and species." *Journal of Human Evolution* 6, no. 1 (1977): 19–53.

Kay, Richard F., Matt Cartmill, and Michelle Balow. "The hypoglossal canal and the origin of human vocal behavior." *Proceedings of the National Academy of Sciences* 95, no. 9 (1998): 5417–5419.

Kay, Richard F., and Herbert H. Covert. "True grit: A microwear experiment." *American Journal of Physical Anthropology* 61, no. 1 (1983): 33–38.

Kiel, Douglas P., Marian T. Hannan, Bruce A. Barton, Mary L. Bouxsein, Thomas F. Lang, Kathleen M. Brown, Elizabeth Shane, Jay Magaziner, Sheryl Zimmerman, and Clinton T. Rubin. "Insights from the conduct of a device trial in older persons: Low magnitude mechanical stimulation for musculoskeletal health." *Clinical Trials* 7, no. 4 (2010): 354–367.

Kimbel, William H., and Lawrence B. Martin, eds. *Species, Species Concepts and Primate Evolution*. New York: Springer Science+Business Media, 2013.

Kimbel, William H., and Yoel Rak. "Functional morphology of the asterionic region in extant hominoids and fossil hominids." *American Journal of Physical Anthropology* 66, no. 1 (1985): 31–54.

Kimbel, William H., and Yoel Rak. "*Australopithecus sediba* and the emergence of *Homo*: Questionable evidence from the cranium of the juvenile holotype MH 1." *Journal of Human Evolution* 107 (2017): 94–106.

King, Mary-Claire, and A. C. Wilson. "Evolution at two levels in humans and chimpanzees." *Science* 188, no. 184 (1975): 107–116.

Kingsolver, Joel G., and Raymond B. Huey. "Introduction: The evolution of morphology, performance, and fitness." *Integrative and Comparative Biology* 43, no. 3 (2003): 361–366.

Kinzey, Warren G. *Evolution of Human Behavior: Primate Models.* Albany: State University of New York Press, 1987.

Kivell, Tracy L., and Daniel Schmitt. "Independent evolution of knuckle-walking in African apes shows that humans did not evolve from a knuckle-walking ancestor." *Proceedings of the National Academy of Sciences* 106, no. 34 (2009): 14241–14246.

Klingenberg, C. P. "Developmental constraints, modules and evolvability" In *Variation: A Central Concept in Biology*, edited by Benedikt Hallgrímsson and Brian K. Hall, 219–247. Burlington, MA: Elsevier, 2005.

Koniecynski, D. D., M. J. Truty, and A. A. Biewener. "Evaluation of a bone's *in vivo* 24-hour loading history for physical exercise compared with background loading." *Journal of Orthopaedic Research* 16, no. 1 (1998): 29–37.

Kotiya, Akhilesh A., Philip V. Bayly, and Matthew J. Silva. "Short-term low-strain vibration enhances chemo-transport yet does not stimulate osteogenic gene expression or cortical bone formation in adult mice." *Bone* 48, no. 3 (2011): 468–475.

Kramer, P. A. "Modelling the locomotor energetics of extinct hominids." *Journal of Experimental Biology* 202, no. 20 (1999): 2807–2818.

Kupczik, K., C. A. Dobson, M. J. Fagan, R. H. Crompton, C. E. Oxnard, and Paul O'Higgins. "Assessing mechanical function of the zygomatic region in macaques: Validation and sensitivity testing of finite element models." *Journal of Anatomy* 210, no. 1 (2007): 41–53.

Lad, Susan E. "The absence of secondary osteons in aged rats." *American Journal of Physical Anthropology* 165, no. S66 (2018): 150.

Lad, Susan E., David J. Daegling, and W. Scott McGraw. "Bone remodeling is reduced in high stress regions of the cercopithecoid mandible." *American Journal of Physical Anthropology* 161, no. 3 (2016): 426–435.

Lague, Michael R. "Another look at shape variation in the distal femur of *Australopithecus afarensis*: Implications for taxonomic and functional diversity at Hadar." *Journal of Human Evolution* 42, no. 5 (2002): 609–626.

LaMothe, Jeremy M., Nicolas H. Hamilton, and Ronald F. Zernicke. "Strain rate influences periosteal adaptation in mature bone." *Medical Engineering & Physics* 27, no. 4 (2005): 277–284.

Langdon, John H. "Umbrella hypotheses and parsimony in human evolution: A critique of the aquatic ape hypothesis." *Journal of Human Evolution* 33, no. 4 (1997): 479–494.

Langdon, John H. *The Human Strategy: An Evolutionary Perspective on Human Anatomy.* Oxford: Oxford University Press, 2005.

Lanyon, Lance E. "Experimental support for the trajectorial theory of bone structure." *Journal of Bone & Joint Surgery: British Volume* 56, no. 1 (1974): 160–166.

Lanyon, Lance E. "Control of bone architecture by functional load bearing." *Journal of Bone and Mineral Research* 7, no. S2 (1992): S369–S375.

Larsen, Clark Spencer. "Biological changes in human populations with agriculture." *Annual Review of Anthropology* 24, no. 1 (1995): 185–213.

Larsen, Clark Spencer. *Bioarchaeology: Interpreting Behavior from the Human Skeleton.* New York: Cambridge University Press, 1997.

Larsen, Clark Spencer. "The agricultural revolution as environmental catastrophe: Implications for health and lifestyle in the Holocene." *Quaternary International* 150, no. 1 (2006): 12–20.

Larson, Susan G. "Rotator cuff muscle size and the interpretation of scapular shape in primates." *Journal of Human Evolution* 80 (2015): 96–106.

Larson, Susan G., and Jack T. Stern Jr. "EMG of scapulohumeral muscles in the chimpanzee during reaching and 'arboreal' locomotion." *American Journal of Anatomy* 176, no. 2 (1986): 171–190.

Latimer, Bruce. "Locomotor adaptations in *Australopithecus afarensis*: The issue of arboreality." In *Origine(s) de la Bipédie chez les Hominidés*, edited by Yves Coppens and Brigitte Senut, 169–176. Cahiers de Paléanthropologie Series. Paris: Éditions du Centre Nationale de la Recherche Scientifique, 1991.

Latimer, Bruce, and C. Owen Lovejoy. "The calcaneus of *Australopithecus afarensis* and its implications for the evolution of bipedality." *American Journal of Physical Anthropology* 78, no. 3 (1989): 369–386.

Latimer, Bruce, and C. Owen Lovejoy. "Hallucal tarsometatarsal joint in *Australopithecus afarensis*." *American Journal of Physical Anthropology* 82, no. 2 (1990a): 125–133.

Latimer, Bruce, and C. Owen Lovejoy. "Metatarsophalangeal joints of *Australopithecus afarensis*." *American Journal of Physical Anthropology* 83, no. 1 (1990b): 13–23.

Latimer, Bruce, James C. Ohman, and C. Owen Lovejoy. "Talocrural joint in African hominoids: Implications for *Australopithecus afarensis*." *American Journal of Physical Anthropology* 74, no. 2 (1987): 155–175.

Lauder, George V. "Form and function: Structural analysis in evolutionary morphology." *Paleobiology* 7, no. 4 (1981): 430–442.

Lauder, George V. "Historical biology and the problem of design." *Journal of Theoretical Biology* 97, no. 1 (1982): 57–67.

Lauder, George V. "Biomechanics and evolution: Integrating physical and historical biology in the study of complex systems." In *Biomechanics in Evolution*, edited by Jeremy M. V. Rayner and Robin J. Wooten, 1–19. Cambridge: Cambridge University Press, 1991.

Lauder, George V. "On the inference of function from structure." In *Functional Morphology in Vertebrate Paleontology*, edited by J. J. Thomason, 1–18. Cambridge: Cambridge University Press, 1995.

Lauder, George V. "The intellectual challenge of biomechanics and evolution." In *Vertebrate Biomechanics and Evolution*, edited by Vincent L. Bells, Jean Pierre Gasc, and Adria Casinos, 319–325. Oxford: BIOS Scientific, 2003.

Lauder, George V., Armand M. Leroi, and Michael R. Rose. "Adaptations and history." *Trends in Ecology and Evolution* 8, no. 8 (1993): 294–297.

Le, Kim N., Matthew Marsik, David J. Daegling, Ana Duque, and William Scott McGraw. "Spatial variation in mandibular bone elastic modulus and its effect on structural bending stiffness: A test case using the Taï Forest monkeys." *American Journal of Physical Anthropology* 162, no. 3 (2017): 516–532.

Leakey, Louis S. B. "A new fossil skull from Olduvai." *Nature* 184, no. 4685 (1959): 491–493.

Leakey, R. E., and A. Walker. "New *Australopithecus boisei* specimens from east and west Lake Turkana, Kenya." *American Journal of Physical Anthropology* 76, no. 1 (1988): 1–24.

Ledogar, Justin A., Amanda L. Smith, Stefano Benazzi, Gerhard W. Weber, Mark A. Spencer, Keely B. Carlson, Kieran P. McNulty, et al. "Mechanical evidence that *Australopithecus sediba* was limited in its ability to eat hard foods." *Nature Communications* 7, no. 1 (2016): 1–9.

Lee, James J.-W., Paul J. Constantino, Peter W. Lucas, and Brian R. Lawn. "Fracture in teeth—a diagnostic for inferring bite force and tooth function." *Biological Reviews* 86, no. 4 (2011): 959–974.

Lee, Na Kyung, Hideaki Sowa, Eiichi Hinoi, Mathieu Ferron, Jong Deok Ahn, Cyrille Confavreux, Romain Dacquin, et al. "Endocrine regulation of energy metabolism by the skeleton." *Cell* 130, no. 3 (2007): 456–469.

Lee, Richard Borshay, and Irven DeVore. *Man the Hunter.* Chicago: Aldine, 1968.

Lee-Thorp, Julia A., Matt Sponheimer, Benjamin H. Passey, Darryl J. de Ruiter, and Thure E. Cerling. "Stable isotopes in fossil hominin tooth enamel suggest a fundamental dietary shift in the Pliocene." *Philosophical Transactions of the Royal Society B: Biological Sciences* 365, no. 1556 (2010): 3389–3396.

Lenaerts, Gerlinde, Friedl De Groote, Bram Demeulenaere, Michiel Mulier, Georges Van der Perre, Arthur Spaepen, and Ilse Jonkers. "Subject-specific hip geometry affects predicted hip joint contact forces during gait." *Journal of Biomechanics* 41, no. 6 (2008): 1243–1252.

Leroi, Armand M., Michael R. Rose, and George V. Lauder. "What does the comparative method reveal about adaptation?" *American Naturalist* 143, no. 3 (1994): 381–402.

Levinton, Jeffrey S. *Genetics, Paleontology, and Macroevolution.* Cambridge: Cambridge University Press, 2001.

Levinton, Jeffrey S., and Chris M. Simon. "A critique of the punctuated equilibria model and implications for the detection of speciation in the fossil record." *Systematic Biology* 29, no. 2 (1980): 130–142.

Li, Jiliang, Randall L. Duncan, David B. Burr, Vincent H. Gattone, and Charles H. Turner. "Parathyroid hormone enhances mechanically induced bone formation, possibly involving L-type voltage-sensitive calcium channels." *Endocrinology* 144, no. 4 (2003): 1226–1233.

Lieberman, Daniel E. "How and why humans grow thin skulls: Experimental evidence for systemic cortical robusticity." *American Journal of Physical Anthropology* 101, no. 2 (1996): 217–236.

Lieberman, Daniel E. "Making behavioral and phylogenetic inferences from hominid fossils: Considering the developmental influence of mechanical forces." *Annual Review of Anthropology* 26, no. 1 (1997): 185–210.

Lieberman, Daniel E. "Ontogeny, homology, and phylogeny in the hominid craniofacial skeleton: The problem of the browridge." In *Development, Growth, and Evolution: Implication for the Study of the Hominid Skeleton*, edited by Paul O'Higgins and Martin Cohn, 85–122. London: Academic Press, 2000.

Lieberman, Daniel E., Osbjorn M. Pearson, John D. Polk, Brigitte Demes, and A. W. Crompton. "Optimization of bone growth and remodeling in response to loading in tapered mammalian limbs." *Journal of Experimental Biology* 206, no. 18 (2003): 3125–3138.

Lieberman, Daniel E., John D. Polk, and Brigitte Demes. "Predicting long bone loading from cross-sectional geometry." *American Journal of Physical Anthropology* 123, no. 2 (2004): 156–171.

Lockwood, Charles A., and John G. Fleagle. "The recognition and evaluation of homoplasy in primate and human evolution." *American Journal of Physical Anthropology* 110, no. S29 (1999): 189–232.

Lovejoy, C. Owen. "Biomechanical perspectives on the lower limb of early hominids." In *Primate Functional Morphology and Evolution*, edited by Russell H. Tuttle, 403–429. Paris: Mouton, 1975.

Lovejoy C. Owen. "A biomechanical review of the locomotor diversity of early hominids." In *Early Hominids of Africa*, edited by Clifford J. Jolly, 403–429. New York: St. Martin's Press, 1978.

Lovejoy, C. Owen. "Reconstruction of the pelvis of AL-288 (Hadar Formation, Ethiopia)." *American Journal of Physical Anthropology* 50, no. 3 (1979): 460.

Lovejoy, C. Owen. "The origin of man." *Science* 211, no. 4480 (1981): 341–350.

Lovejoy, C. Owen. "Evolution of human walking." *Scientific American* 259, no. 5 (1988): 118–125.

Lovejoy, C. Owen. "Reexamining human origins in light of *Ardipithecus ramidus*." *Science* 326, no. 5949 (2009): 74–74e8.

Lovejoy, C. Owen, Martin J. Cohn, and Tim D. White. "Morphological analysis of the mammalian postcranium: A developmental perspective." *Proceedings of the National Academy of Sciences* 96, no. 23 (1999): 13247–13252.

Lovejoy, C. Owen, Kingsbury G. Heiple, and Albert H. Burstein. "The gait of *Australopithecus*." *American Journal of Physical Anthropology* 38, no. 3 (1973): 757–779.

Lovejoy, C. Owen, Donald C. Johanson, and Yves Coppens. "Hominid lower limb bones recovered from the Hadar Formation: 1974–1977 collections." *American Journal of Physical Anthropology* 57, no. 4 (1982): 679–700.

Lovejoy, C. Owen, Richard S. Meindl, James C. Ohman, Kingsbury G. Heiple, and Tim D. White. "The Maka femur and its bearing on the antiquity of human walking: Applying contemporary concepts of morphogenesis to the human fossil record." *American Journal of Physical Anthropology* 119, no. 2 (2002): 97–133.

Lovejoy, C. Owen, Gen Suwa, Scott W. Simpson, Jay H. Matternes, and Tim D. White. "The great divides: *Ardipithecus ramidus* reveals the postcrania of our last common ancestors with African apes." *Science* 326, no. 5949 (2009): 73–106.

Lovell, Nancy C. "An evolutionary framework for assessing illness and injury in nonhuman primates." *American Journal of Physical Anthropology* 34, no. S13 (1991): 117–155.

Lucas, Peter [W.], Paul Constantino, Bernard Wood, and Brian Lawn. "Dental enamel as a dietary indicator in mammals." *BioEssays* 30, no. 4 (2008): 374–385.

Lucas, Peter W., Ridwaan Omar, Khaled Al-Fadhalah, Abdulwahab S. Almusallam, Amanda G. Henry, Shaji Michael, Lidia Arockia Thai, Jörg Watzke, David S. Strait, and Anthony G. Atkins. "Mechanisms and causes of wear in tooth enamel: Implications for hominin diets." *Journal of the Royal Society Interface* 10, no. 80 (2013): article ID 20120923.

Lucas, Peter W., Adam van Casteren, Khaled Al-Fadhalah, Abdulwahab S. Almusallam, Amanda G. Henry, Shaji Michael, Jörg Watzke, et al. "The role of dust, grit and phytoliths in tooth wear." In *Annales Zoologici Fennici* 51, no. 1 (2014): 143–152.

Macho, Gabriele A. "General principles of evolutionary morphology" In *Handbook of Paleoanthropology*, edited by Winfried Henke and Ian Tattersall, 921–936. Heidelberg: Springer, 2013.

Madar, Sandra I., Michael D. Rose, Jay Kelley, Laura MacLatchy, and David Pilbeam. "New *Sivapithecus* postcranial specimens from the Siwaliks of Pakistan." *Journal of Human Evolution* 42, no. 6 (2002): 705–752.

Manafzadeh, Armita R., and Kevin Padian. "ROM mapping of ligamentous constraints on avian hip mobility: Implications for extinct ornithodirans." *Proceedings of the Royal Society B: Biological Sciences* 285, no. 1879 (2018): 20180727.

Marks, Jonathan. "Genetic assimilation in the evolution of bipedalism." *Human Evolution* 4, no. 6 (1989): 493–499.

Marks, Jonathan. "Phylogenetic trees and evolutionary forests." *Evolutionary Anthropology: Issues, News, and Reviews* 14, no. 2 (2005): 49–53.

Marroig, Gabriel, and James M. Cheverud. "A comparison of phenotypic variation and covariation patterns and the role of phylogeny, ecology, and ontogeny during cranial evolution of New World monkeys." *Evolution* 55, no. 12 (2001): 2576–2600.

Martin, R[obert] B[ruce]. "Is all cortical bone remodeling initiated by microdamage?" *Bone* 30, no. 1 (2002): 8–13.

Martin, R[obert] Bruce, David B. Burr, Neil A. Sharkey, and David P. Fyhrie. *Skeletal Tissue Mechanics*. New York: Springer, 2015.

Martin, Robert D. "Towards a new definition of primates." *Man* 3, no. 3 (1968): 377–401.

Maslow, Abraham H. *The Psychology of Science: A Reconnaissance*. Chicago: Henry Regnery (1966).

Mayr, Ernst. *Animal Species and Evolution*, Cambridge, MA: Harvard University Press, 1963.

Mayr, Ernst. "How to carry out the adaptationist program?" *American Naturalist* 121, no. 3 (1983): 324–334.

McBirney, Alex, and Stanton Cook, trans. and annot. *The Philosophy of Zoology before Darwin: A Translated and Annotated Version of the Original French Text by Edmond Perrier*. Originally published Paris: Félix Alcan, 1884. Translation published Dordrecht, Netherlands: Springer Science+Business Media, 2009.

McCollum, Melanie A. "Palatal thickening and facial form in *Paranthropus*: Examination of alternative developmental models." *American Journal of Physical Anthropology* 103, no. 3 (1997): 375–392.

McCollum, Melanie A. "The robust australopithecine face: A morphogenetic perspective." *Science* 284, no. 5412 (1999): 301–305.

McCormack, Steven W., Ulrich Witzel, Peter J. Watson, Michael J. Fagan, and Flora Gröning. "Inclusion of periodontal ligament fibres in mandibular finite element models leads to an increase in alveolar bone strains." *PLoS One* 12, no. 11 (2017): e0188707.

McGhee, George R. *Theoretical Morphology: The Concept and Its Applications*. New York: Columbia University Press, 1999.

McGraw, W. Scott, and David J. Daegling. "Diet, feeding behavior, and jaw architecture of Taï monkeys: Congruence and chaos in the realm of functional morphology." *Evolutionary Anthropology: Issues, News, and Reviews* 29, no. 1 (2020): 14–28.

McGraw, W. Scott, James D. Pampush, and David J. Daegling. "Brief communication: Enamel thickness and durophagy in mangabeys revisited." *American Journal of Physical Anthropology* 147, no. 2 (2012): 326–333.

McGraw, W. Scott, Klaus Zuberbühler, and Ronald Noë, eds. *Monkeys of the Taï Forest: An African Primate Community*. Cambridge Studies in Biological and Evolutionary Anthropology No. 51. Cambridge: Cambridge University Press, 2007.

McHenry, Henry M. "New estimates of body weight in early hominids and their significance to encephalization and megadontia in 'robust' australopithecines." In *Evolutionary History of the "Robust" Australopithecines*, edited by Frederick E. Grine, 133–148. New York: Aldine de Gruyter, 1988.

McHenry, Henry M. "Petite bodies of the 'robust' australopithecines." *American Journal of Physical Anthropology* 86, no. 4 (1991): 445–454.

McHenry, Henry M. "Body size and proportions in early hominids." *American Journal of Physical Anthropology* 87, no. 4 (1992): 407–431.

McLeod, Kenneth J., and Clinton T. Rubin. "The effect of low-frequency electrical fields on osteogenesis." *Journal of Bone & Joint Surgery* 74, no. 6 (1992): 920–929.

Mednikova, M. B., M. V. Dobrovolskaya, B. Viola, A. V. Lavrenyuk, P. R. Kazansky, V. Y. Shklover, M. V. Shunkov, and A. P. Derevianko. "A micro computerized tomography (X-ray microscopy) of the hand phalanx of the Denisova girl." *Archaeology, Ethnology and Anthropology of Eurasia* 41, no. 3 (2013): 120–125.

Meldrum, D. Jeffrey. "Fossilized Hawaiian footprints compared with Laetoli hominid footprints." In *From Biped to Strider: The Emergence of Modern Human Walking, Running, and Resource Transport*, edited by D. Jeffrey Meldrum and Charles E. Hilton, 63–83. Boston: Springer, 2004.

Meldrum, D. Jeffrey, and Charles E. Hilton, eds. *From Biped to Strider: The Emergence of Modern Human Walking, Running, and Resource Transport*. Boston: Springer, 2004.

Menegaz, Rachel A., and Matthew J. Ravosa. "Ontogenetic and functional modularity in the rodent mandible." *Zoology* 124 (2017): 61–72.

Menegaz, Rachel A., Samantha V. Sublett, Said D. Figueroa, Timothy J. Hoffman, Matthew J. Ravosa, and Kristina Aldridge. "Evidence for the influence of diet on cranial form and robusticity." *Anatomical Record: Advances in Integrative Anatomy and Evolutionary Biology* 293, no. 4 (2010): 630–641.

Miles, Donald B. "The race goes to the swift: Fitness consequences of variation in sprint performance in juvenile lizards." *Evolutionary Ecology Research* 6, no. 1 (2004): 63–75.

Millette, James B., Michelle L. Sauther, and Frank P. Cuozzo. "Behavioral responses to tooth loss in wild ring-tailed lemurs (*Lemur catta*) at the Beza Mahafaly Special Reserve, Madagascar." *American Journal of Physical Anthropology* 40, no. 1 (2009): 120–134.

Montagu, M. Ashley. "Time, morphology, and neoteny in the evolution of man." *American Anthropologist* 57, no. 1 (1955): 13–27.

Mori, S., and D. B. Burr. "Increased intracortical remodeling following fatigue damage." *Bone* 14, no. 2 (1993): 103–109.

Morris, Desmond. *The Naked Ape: A Zoologist's Study of the Human Animal.* New York: Random House, 1967.

Moss, Melvin L. "A theoretical analysis of the functional matrix." *Acta Biotheoretica* 18, no. 1–4 (1968): 195–202.

Moss, Melvin L. "A functional cranial analysis of primate craniofacial growth." In *Craniofacial Biology of Primates*, edited by M. R. Zingeser, 191–208. Vol. 3 of *Symposium of the Fourth International Congress of Primatology*. Basel, Switzerland: S. Karger, 1973.

Moss, Melvin L. "The functional matrix hypothesis revisited, 1: The role of mechano-transduction." *American Journal of Orthodontics and Dentofacial Orthopedics* 112, no. 1 (1997a): 8–11.

Moss, Melvin L. "The functional matrix hypothesis revisited, 2: The role of an osseous connected cellular network." *American Journal of Orthodontics and Dentofacial Orthopedics* 112, no. 2 (1997b): 221–226.

Moss, Melvin L. "The functional matrix hypothesis revisited, 3: The genomic thesis." *American Journal of Orthodontics and Dentofacial Orthopedics* 112, no. 3 (1997c): 338–342.

Moss, Melvin L. "The functional matrix hypothesis revisited, 4: The epigenetic antithesis and the resolving synthesis." *American Journal of Orthodontics and Dentofacial Orthopedics* 112, no. 4 (1997d): 410–417.

Moss, Melvin L., and Richard W. Young. "A functional approach to craniology." *American Journal of Physical Anthropology* 18, no. 4 (1960): 281–292.

Müller, Gerd B., and Günter P. Wagner. "Novelty in evolution: Restructuring the concept." *Annual Review of Ecology and Systematics* 22, no. 1 (1991): 229–256.

Nakai, M. "Bone and joint disorders in wild Japanese macaques from Nagano Prefecture, Japan." *International Journal of Primatology* 24, no. 1 (2003): 179–195.

Nakajima, K., M. A. Maier, P. A. Kirkwood, and R. N. Lemon. "Striking differences in transmission of corticospinal excitation to upper limb motoneurons in two primate species." *Journal of Neurophysiology* 84, no. 2 (2000): 698–709.

Nakatsukasa, Masato, Sugio Hayama, and Holger Preuschoft. "Postcranial skeleton of a macaque trained for bipedal standing and walking and implications for functional adaptation." *Folia Primatologica* 64, no. 1–2 (1995): 1–29.

Nemeschkal, H. L., R. van den Elzen, and H. Brieschke. "The morphometric extraction of character complexes accomplishing common biological roles: Avian skeletons as a case study." *Journal of Zoological Systematics and Evolutionary Research* 30, no. 3 (1992): 201–219.

Nordström, A., C. Karlsson, F. Nyquist, T. Olsson, P. Nordström, and M. Karlsson. "Bone loss and fracture risk after reduced physical activity." *Journal of Bone and Mineral Research* 20, no. 2 (2005): 202–207.

Nowlan, Niamh C., and Patrick J. Prendergast. "Evolution of mechanoregulation of bone growth will lead to non-optimal bone phenotypes." *Journal of Theoretical Biology* 235, no. 3 (2005): 408–418.

Nunn, Charles L. *The Comparative Approach in Evolutionary Anthropology and Biology.* Chicago: University of Chicago Press, 2011.

O'Grady, Richard T. "Evolutionary theory and teleology." *Journal of Theoretical Biology* 107, no. 4 (1984): 563–578.

Ohman, James C. "The first rib of hominoids." *American Journal of Physical Anthropology* 70, no. 2 (1986): 209–229.

Ohman, James C., Todd J. Krochta, C. Owen Lovejoy, Robert P. Mensforth, and Bruce Latimer. "Cortical bone distribution in the femoral neck of hominoids: Implications for the locomotion of *Australopithecus afarensis*." *American Journal of Physical Anthropology* 104, no. 1 (1997): 117–131.

Oxnard, Charles E[rnest]. "The functional morphology of the primate shoulder as revealed by comparative anatomical, osteometric and discriminant function techniques." *American Journal of Physical Anthropology* 26, no. 2 (1967): 219–240.

Oxnard, Charles E[rnest]. *Form and Pattern in Human Evolution: Some Mathematical, Physical, and Engineering Approaches.* Chicago: University of Chicago Press, 1973a.

Oxnard, Charles E[rnest]. "Functional inferences from morphometrics: Problems posed by uniqueness and diversity among the primates." *Systematic Zoology* 22, no. 4 (1973b): 409–424.

Oxnard, Charles E[rnest]. *Uniqueness and Diversity in Human Evolution: Morphometric Studies of Australopithecines.* Chicago: University of Chicago Press, 1975.

Oxnard, Charles E[rnest]. "One biologist's view of morphometrics." *Annual Review of Ecology and Systematics* 9, no. 1 (1978): 219–241.

Oxnard, Charles Ernest. "Introduction to the symposium: Analysis of form; Some problems underlying most studies of form." *American Zoologist* 20, no. 4 (1980): 619–626.

Oxnard, Charles Ernest. *The Order of Man: A Biomathematical Anatomy of the Primates.* New Haven, CT: Yale University Press, 1984.

Oxnard, Charles Ernest. "Design, level, interface, and complexity: Morphometric interpretation revisited." In *Shaping Primate Evolution: Form, Function, and Behavior,* edited by Fred Anapol, Rebecca Z. German, and Nina G. Jablonski, 391–414. Cambridge Studies in Biological and Evolutionary Anthropology No. 40. Cambridge: Cambridge University Press, 2004.

Ozcivici, Engin, Yen Kim Luu, Ben Adler, Yi-Xian Qin, Janet Rubin, Stefan Judex, and Clinton T. Rubin. "Mechanical signals as anabolic agents in bone." *Nature Reviews: Rheumatology* 6, no. 1 (2010): 50–59.

Padian, Kevin. "Form versus function: The evolution of a dialectic." In *Functional Morphology in Vertebrate Paleontology,* edited by J. J. Thomason, 264–277. Cambridge: Cambridge University Press, 1995.

Pampush, James D. "Selection played a role in the evolution of the human chin." *Journal of Human Evolution* 82 (2015): 127–136.

Pampush, James D., and David J. Daegling. "The enduring puzzle of the human chin." *Evolutionary Anthropology: Issues, News, and Reviews* 25, no. 1 (2016): 20–35.

Pampush, James D., Ana C. Duque, Brittany R. Burrows, David J. Daegling, William F. Kenney, and W. Scott McGraw. "Homoplasy and thick enamel in primates." *Journal of Human Evolution* 64, no. 3 (2013): 216–224.

Parfitt, A. Michael. "Osteonal and hemi-osteonal remodeling: The spatial and temporal framework for signal traffic in adult human bone." *Journal of Cellular Biochemistry* 55, no. 3 (1994): 273–286.

Parfitt, A. Michael. "Targeted and nontargeted bone remodeling: Relationship to basic multicellular unit origination and progression." *Bone* 30, no. 1 (2002): 5–7.

Parins-Fukuchi, Caroline. "Mosaic evolution, preadaptation, and the evolution of evolvability in apes." *Evolution* 74, no. 2 (2020): 297–310.

Pauwels, Friedrich. *Biomechanics of the Locomotor Apparatus.* Berlin: Springer, 1980.

Pearson, Alannah, Colin Groves, and Andrea Cardini. "The 'temporal effect' in hominids: Reinvestigating the nature of support for a chimp-human clade in bone morphology." *Journal of Human Evolution* 88 (2015): 146–159.

Pearson, Osbjorn M., and Daniel E. Lieberman. "The aging of Wolff's 'law': Ontogeny and responses to mechanical loading in cortical bone." *American Journal of Physical Anthropology* 125, no. S39 (2004): 63–99.

Percival, Christopher J., Rebecca Green, Charles C. Roseman, Daniel M. Gatti, Judith L. Morgan, Stephen A. Murray, Leah Rae Donahue, et al. "Developmental constraint through negative pleiotropy in the zygomatic arch." *EvoDevo* 9, no. 1 (2018): 1–16.

Perry, Jonathan M. G., and Christine E. Wall. "Scaling of the chewing muscles in prosimians." In *Primate Craniofacial Function and Biology*, edited by Christopher Vinyard, Matthew J. Ravosa, and Christine E. Wall, 217–240. Boston: Springer, 2008.

Peters, Charles R. "Nut-like oil seeds: Food for monkeys, chimpanzees, humans, and probably ape-men." *American Journal of Physical Anthropology* 73, no. 3 (1987): 333–363.

Peters, Dieter Stefan. "Behavior plus 'pathology'—the origin of adaptations?" In *Constructional Morphology and Evolution*, edited by Norbert Schmidt-Kittler and Klaus Vogel, 141–150. Berlin: Springer, 1991.

Peterson, Jane. *Sexual Revolutions: Gender and Labor at the Dawn of Agriculture.* Walnut Creek, CA: AltaMira, 2002.

Pigliucci, Massimo. *Phenotypic Plasticity: Beyond Nature and Nurture.* Baltimore: Johns Hopkins University Press, 2001.

Pigliucci, Massimo. "Phenotypic integration: Studying the ecology and evolution of complex phenotypes." *Ecology Letters* 6, no. 3 (2003): 265–272.

Pilbeam, David [R.], and Stephen Jay Gould. "Size and scaling in human evolution." *Science* 186, no. 4167 (1974): 892–901.

Pilbeam, David R., and Daniel E. Lieberman. "Reconstructing the last common ancestor of chimpanzees and humans." *Chimpanzees and Human Evolution*, edited by Martin N. Muller, 22-141. Cambridge, MA: Harvard University Press, 2017.

Pivonka, Peter, Aaron Park, and Mark R. Forwood. "Functional adaptation of bone: The mechanostat and beyond." In *Multiscale Mechanobiology of Bone Remodeling and Adaptation*, edited by Peter Pivonka, 1–60. Cham, Switzerland: Springer, 2018.

Plavcan, J. Michael, and David J. Daegling. "Interspecific and intraspecific relationships between tooth size and jaw size in primates." *Journal of Human Evolution* 51, no. 2 (2006): 171–184.

Plotnick, Roy E., and Tomasz K. Baumiller. "Invention by evolution: Functional analysis in paleobiology." *Paleobiology* 26, no. 4 (2000): 305–323.

Pontzer, Herman. "The crown joules: Energetics, ecology, and evolution in humans and other primates." *Evolutionary Anthropology: Issues, News, and Reviews* 26, no. 1 (2017a): 12–24.

Pontzer, Herman. "Economy and endurance in human evolution." *Current Biology* 27, no. 12 (2017b): R613–R621.

Pontzer, Herman, David A. Raichlen, and Peter S. Rodman. "Bipedal and quadrupedal locomotion in chimpanzees." *Journal of Human Evolution* 66 (2014): 64–82.

Potts, Richard. "Evolution and climate variability." *Science* 273, no. 5277 (1996): 922–923.

Potts, Richard. "Variability selection in hominid evolution." *Evolutionary Anthropology: Issues, News, and Reviews* 7, no. 3 (1998): 81–96.

Pross, Addy. "On the chemical nature and origin of teleonomy." *Origins of Life and Evolution of Biospheres* 35, no. 4 (2005): 383–394.

Qin, Yi-Xian, Clinton T. Rubin, and Kenneth J. McLeod. "Nonlinear dependence of loading intensity and cycle number in the maintenance of bone mass and morphology." *Journal of Orthopaedic Research* 16, no. 4 (1998): 482–489.

Rabenold, Diana, and Osbjorn M. Pearson. "Abrasive, silica phytoliths and the evolution of thick molar enamel in primates, with implications for the diet of *Paranthropus boisei*." *PLoS One* 6, no. 12 (2011): e28379.

Radinsky, L. B. "Approaches in evolutionary morphology: A search for patterns." *Annual Review of Ecology and Systematics* 16, no. 1 (1985): 1–14

Rafferty, Katherine L. "Structural design of the femoral neck in primates." *Journal of Human Evolution* 34, no. 4 (1998): 361–383.

Raichlen, David A., and Adam D. Gordon. "Interpretation of footprints from Site S confirms human-like bipedal biomechanics in Laetoli hominins." *Journal of Human Evolution* 107 (2017): 134–138.

Raichlen, David A., Adam D. Gordon, William E. H. Harcourt-Smith, Adam D. Foster, and Wm. Randall Haas Jr. "Laetoli footprints preserve earliest direct evidence of human-like bipedal biomechanics." *PLoS One* 5, no. 3 (2010): e9769.

Rak, Yoel. *The Australopithecine Face*. New York: Academic Press, 1983.

Ramdarshan, A., T. Alloing-Séguier, G. Merceron, and L. Marivaux. "The primate community of Cachoeira (Brazilian Amazonia): A model to decipher ecological partitioning among extinct species." *PLoS One* 6, no. 11 (2011): e27392.

Rapoff, Andrew J., Renaud G. Rinaldi, Jennifer L. Hotzman, and David J. Daegling. "Elastic modulus variation in mandibular bone: A microindentation study of *Macaca fascicularis*." *American Journal of Physical Anthropology* 135, no. 1 (2008): 100–109.

Raup, David M. "Geometric analysis of shell coiling: General problems." *Journal of Paleontology* (1966): 1178–1190.

Raup, David M. "Approaches to morphologic analysis." In *Models in Paleobiology*, edited by Thomas J. M. Schopf, 28–44. San Francisco: Freeman, Cooper, 1972.

Raup, David M., and Arnold Michelson. "Theoretical morphology of the coiled shell." *Science* 147, no. 3663 (1965): 1294–1295.

Ravosa, Matthew J. "Browridge development in Cercopithecidae: A test of two models." *American Journal of Physical Anthropology* 76, no. 4 (1988): 535–555.

Ravosa, Matthew J. "Ontogenetic perspective on mechanical and nonmechanical models of primate circumorbital morphology." *American Journal of Physical Anthropology* 85, no. 1 (1991): 95–112.

Ravosa, Matthew J. "Jaw morphology and function in living and fossil Old World monkeys." *International Journal of Primatology* 17, no. 6 (1996): 909–932.

Ravosa, Matthew J. "Anthropoid origins and the modern symphysis." *Folia Primatologica* 70, no. 2 (1999): 65–78.

Ravosa, Matthew J. "Size and scaling in the mandible of living and extinct apes." *Folia Primatologica* 71, no. 5 (2000): 305–322.

Ravosa, Matthew J., Kirk R. Johnson, and William L. Hylander. "Strain in the Galago facial skull." *Journal of Morphology* 245, no. 1 (2000): 51–66.

Ravosa, Matthew J., and Robert J. Kane. "Dietary variation and mechanical properties of articular cartilage in the temporomandibular joint: Implications for the role of plasticity in mechanobiology and pathobiology." *Zoology* 124 (2017): 42–50.

Ravosa, Matthew J., Ravinder Kunwar, Stuart R. Stock, and M. Sharon Stack. "Pushing the limit: Masticatory stress and adaptive plasticity in mammalian craniomandibular joints." *Journal of Experimental Biology* 210, no. 4 (2007): 628–641.

Ravosa, Matthew J., Rachel A. Menegaz, Jeremiah E. Scott, David J. Daegling, and Kevin R. McAbee. "Limitations of a morphological criterion of adaptive inference in the fossil record." *Biological Reviews* 91, no. 4 (2016): 883–898.

Ravosa, Matthew J., Callum F. Ross, Susan H. Williams, and Destiny B. Costley. "Allometry of masticatory loading parameters in mammals." *Anatomical Record: Advances in Integrative Anatomy and Evolutionary Biology* 293, no. 4 (2010): 557–571.

Ravosa, Matthew J., and Christopher J. Vinyard. "On the interface between ontogeny and function." In *Reconstructing Behavior in the Primate Fossil Record*, edited by J. Michael Plavcan, Richard F. Kay, William L. Jungers, and Carel P. van Schaik, 43–112. New York: Kluwer Academic, 2002.

Rayfield, Emily J. "Finite element analysis and understanding the biomechanics and evolution of living and fossil organisms." *Annual Review of Earth and Planetary Sciences* 35, no. 1 (2007): 541–576.

Reed, Kaye E. "Early hominid evolution and ecological change through the African Plio-Pleistocene." *Journal of Human Evolution* 32, no. 2–3 (1997): 289–322.

Reeve, Hudson Kern, and Paul W. Sherman. "Adaptation and the goals of evolutionary research." *Quarterly Review of Biology* 68, no. 1 (1993): 1–32.

Reif, Wolf-Ernst, Roger D. K. Thomas, and Martin S. Fischer. "Constructional morphology: The analysis of constraints in evolution." *Acta Biotheoretica* 34, no. 2–4 (1985): 233–248.

Rein, Thomas R. "The correspondence between proximal phalanx morphology and locomotion: Implications for inferring the locomotor behavior of fossil catarrhines." *American Journal of Physical Anthropology* 146, no. 3 (2011): 435–445.

Richmond, Brian G. "Biomechanics of phalangeal curvature." *Journal of Human Evolution* 53, no. 6 (2007): 678–690.

Richmond, Brian G., Leslie C. Aiello, and Bernard A. Wood. "Early hominin limb proportions." *Journal of Human Evolution* 43 (2002): 529–548.

Richmond, Brian G., David R. Begun, and David S. Strait. "Origin of human bipedalism: The knuckle-walking hypothesis revisited." *Yearbook of Physical Anthropology* 44, no. S33 (2001): 70–105.

Richmond, Brian G., and David S. Strait. "Evidence that humans evolved from a knuckle-walking ancestor." *Nature* 404, no. 6776 (2000): 382–385.

Richmond, Brian G., and David S. Strait. "Knuckle-walking hominid ancestor: A reply to Corruccini & McHenry." *Journal of Human Evolution* 40, no. 6 (2001): 513–520.

Riddle, Ryan C., and Henry J. Donahue. "From streaming-potentials to shear stress: 25 years of bone cell mechanotransduction." *Journal of Orthopaedic Research* 27, no. 2 (2009): 143–149.

Rieppel, Olivier. "Structuralism, functionalism, and the four Aristotelian causes." *Journal of the History of Biology* 23, no. 2 (1990): 291–320.

Rizzoli, René. "Nutritional aspects of bone health." *Best Practice & Research: Clinical Endocrinology & Metabolism* 28, no. 6 (2014): 795–808.

Robinson, John T. "Prehominid dentition and hominid evolution." *Evolution* 8, no. 4 (1954): 324–334.

Robling, Alexander G. "Is bone's response to mechanical signals dominated by muscle forces?" *Medicine and Science in Sports and Exercise* 41, no. 11 (2009): 2044.

Robling, A[lexander] G., K. M. Duijvelaar, J. V. Geevers, N. Ohashi, and C. H. Turner. "Modulation of appositional and longitudinal bone growth in the rat ulna by applied static and dynamic force." *Bone* 29, no. 2 (2001): 105–113.

Robling, Alexander G., Felicia M. Hinant, David B. Burr, and Charles H. Turner. "Improved bone structure and strength after long-term mechanical loading is greatest if loading is separated into short bouts." *Journal of Bone and Mineral Research* 17, no. 8 (2002): 1545–1554.

Robling, Alexander G., and Charles H. Turner. "Mechanotransduction in bone: Genetic effects on mechanosensitivity in mice." *Bone* 31, no. 5 (2002): 562–569.

Roll-Hansen, Nils. "E. S. Russell and J. H. Woodger: The failure of two twentieth-century opponents of mechanistic biology." *Journal of the History of Biology* 17, no. 3 (1984): 399–428.

Rose, Michael R., and George V. Lauder. "Post-spandrel adaptationism." In *Adaptation*, edited by Michael R. Rose and George V. Lauder, 1–8. San Diego: Academic Press, 1996.

Ross, Callum F., Ruchi Dharia, Susan W. Herring, William L. Hylander, Zi-Jun Liu, Katherine L. Rafferty, Matthew J. Ravosa, and Susan H. Williams. "Modulation of mandibular loading and bite force in mammals during mastication." *Journal of Experimental Biology* 210, no. 6 (2007): 1046–1063.

Ross, Callum F., Jose Iriarte-Diaz, and Charles L. Nunn. "Innovative approaches to the relationship between diet and mandibular morphology in primates." *International Journal of Primatology* 33, no. 3 (2012): 632–660.

Ross, Callum F., Jose Iriarte-Diaz, David A. Reed, Thomas A. Stewart, and Andrea B. Taylor. "*In vivo* bone strain in the mandibular corpus of *Sapajus* during a range of oral food processing behaviors." *Journal of Human Evolution* 98 (2016): 36–65.

Ross, Callum F., Charles A. Lockwood, John G. Fleagle, and William L. Jungers. "Adaptation and behavior in the primate fossil record." In *Reconstructing Behavior in the Primate Fossil Record*, edited by J. Michael Plavcan, Richard F. Kay, William L. Jungers, and Carel P. van Schaik, 1–41. Boston: Springer, 2002.

Ross, Callum F., Biren A. Patel, Dennis E. Slice, David S. Strait, Paul C. Dechow, Brian G. Richmond, and Mark A. Spencer. "Modeling masticatory muscle force in finite element analysis: Sensitivity analysis using principal coordinates analysis." *Anatomical Record Part A: Discoveries in Molecular, Cellular, and Evolutionary Biology* 283, no. 2 (2005): 288–299.

Ross, Caroline. "The intrinsic rate of natural increase and reproductive effort in primates." *Journal of Zoology* 214, no. 2 (1988): 199–219.

Ross, Caroline. "Basal metabolic rate, body weight and diet in primates: An evaluation of the evidence." *Folia Primatologica* 58, no. 1 (1992): 7–23.

Rubin, Clinton [T.], Stefan Judex, and Yi-Xian Qin. "Low-level mechanical signals and their potential as a non-pharmacological intervention for osteoporosis." *Age and Ageing* 35, Suppl. 2 (2006): ii32–ii36.

Rubin, Clinton T., and Lance E. Lanyon. "Dynamic strain similarity in vertebrates; An alternative to allometric limb bone scaling." *Journal of Theoretical Biology* 107, no. 2 (1984a): 321–327.

Rubin, Clinton T., and Lance E. Lanyon. "Regulation of bone formation by applied dynamic loads." *Journal of Bone & Joint Surgery: American Volume* 66 (1984b): 397–402.

Rubin, Clinton T., and Lance E. Lanyon. "Regulation of bone mass by mechanical strain magnitude." *Calcified Tissue International* 37, no. 4 (1985): 411–417.

Rubin, Clinton T., and Lance E. Lanyon. "Osteoregulatory nature of mechanical stimuli: Function as a determinant for adaptive remodeling in bone." *Journal of Orthopaedic Research* 5, no. 2 (1987): 300–310.

Rubin, Clinton T., Kevin J. McLeod, Ted S. Gross, and Henry J. Donahue "Physical stimuli as potent determinants of bone morphology." In *Bone Biodynamics in Orthodontic and Orthopedic Treatment*, edited by David S. Carlson, and Steven A. Goldstein, 75–91. Craniofacial Growth Series No. 27. Ann Arbor: Center for Human Growth and Development, University of Michigan, 1992.

Rubin, Clinton [T.], Robert Recker, Diane Cullen, John Ryaby, Joan McCabe, and Kenneth McLeod. "Prevention of postmenopausal bone loss by a low-magnitude, high-frequency mechanical stimuli: A clinical trial assessing compliance, efficacy, and safety." *Journal of Bone and Mineral Research* 19, no. 3 (2004): 343–351.

Rubin, Clinton [T.], A. Simon Turner, Steven Bain, Craig Mallinckrodt, and Kenneth McLeod. "Low mechanical signals strengthen long bones." *Nature* 412, no. 6847 (2001): 603–604.

Rubin, C[linton T.], A. S. Turner, C. Mallinckrodt, C. Jerome, K. McLeod, and S. Bain. "Mechanical strain, induced noninvasively in the high-frequency domain, is anabolic to cancellous bone, but not cortical bone." *Bone* 30, no. 3 (2002): 445–452.

Rudwick, Martin John Spencer. "The inference of function from structure in fossils." *British Journal for the Philosophy of Science* 15, no. 57 (1964): 27–40.

Ruff, Christopher B. "Body size, body shape, and long bone strength in modern humans." *Journal of Human Evolution* 38, no. 2 (2000): 269–290.

Ruff, Christopher B. "Gracilization of the modern human skeleton: The latent strength in our slender bones teaches lessons about human lives, current and past." *American Scientist* 94, no. 6 (2006): 508–514.

Ruff, Christopher, Brigitte Holt, and Erik Trinkaus. "Who's afraid of the big bad Wolff? 'Wolff's law' and bone functional adaptation." *American Journal of Physical Anthropology* 129, no. 4 (2006): 484–498.

Ruff, Christopher B., Clark Spencer Larsen, and Wilson C. Hayes. "Structural changes in the femur with the transition to agriculture on the Georgia coast." *American Journal of Physical Anthropology* 64, no. 2 (1984): 125–136.

Ruff, Christopher B., Henry M. McHenry, and J. Francis Thackeray. "Cross-sectional morphology of the SK 82 and 97 proximal femora." *American Journal of Physical Anthropology* 109, no. 4 (1999): 509–521.

Ruff, C[hristopher] B., and J. A. Runestad. "Primate limb bone structural adaptations." *Annual Review of Anthropology* 21, no. 1 (1992): 407–433.

Ruse, Michael. "Scientific creationism." In *Darwinism Defended: A Guide to the Evolution Controversies*, 293–302. Menlo Park, CA: Benjamin Cummings, 1982.

Ruse, Michael. "Teleology: Yesterday, today, and tomorrow?" *Studies in History and Philosophy of Science Part C: Studies in History and Philosophy of Biological and Biomedical Sciences* 31, no. 1 (2000): 213–232.

Russell, Edward Stuart. *Form and Function: A Contribution to the History of Animal Morphology*. London: John Murray, 1916.

Russell, Mary Doria. "The supraorbital torus: A most remarkable peculiarity." *Current Anthropology* 26, no. 3 (1985): 337–360.

Russo, Gabrielle A., and E. Christopher Kirk. "Foramen magnum position in bipedal mammals." *Journal of Human Evolution* 65, no. 5 (2013): 656–670.

Russo, Gabrielle A., D'arcy Marsh, and Adam D. Foster. "Response of the axial skeleton to bipedal loading behaviors in an experimental animal model." *Anatomical Record: Advances in Integrative Anatomy and Evolutionary Biology* 303, no. 1 (2018): 150–166.

Ruxton, Graeme D., and David M. Wilkinson. "Avoidance of overheating and selection for both hair loss and bipedality in hominins." *Proceedings of the National Academy of Sciences* 108, no. 52 (2011): 20965–20969.

Ryan, Timothy M., Kristian J. Carlson, Adam D. Gordon, Nina Jablonski, Colin N. Shaw, and Jay T. Stock. "Human-like hip joint loading in *Australopithecus africanus* and *Paranthropus robustus*." *Journal of Human Evolution* 121 (2018): 12–24.

Ryan, Timothy M., and Richard A. Ketcham. "Femoral head trabecular bone structure in two omomyid primates." *Journal of Human Evolution* 43, no. 2 (2002): 241–263.

Sanchez, Sophie, Per E. Ahlberg, Katherine M. Trinajstic, Alessandro Mirone, and Paul Tafforeau. "Three-dimensional synchrotron virtual paleohistology: A new insight into the world of fossil bone microstructures." *Microscopy and Microanalysis* 18, no. 5 (2012): 1095–1105.

Santos, Ana, Astrid D. Bakker, and Jenneke Klein-Nulend. "The role of osteocytes in bone mechanotransduction." *Osteoporosis International* 20, no. 6 (2009): 1027–1031.

Schaeffer, Bobb. "The role of experimentation in the origin of higher levels of organization." *Systematic Zoology* 14, no. 4 (1965): 318–336.

Schaller, George B., and Gordon R. Lowther. "The relevance of carnivore behavior to the study of early hominids." *Southwestern Journal of Anthropology* 25, no. 4 (1969): 307–341.

Schmitt, Daniel. "Compliant walking in primates." *Journal of Zoology* 248, no. 2 (1999): 149–160.

Schmitt, Daniel. "Insights into the evolution of human bipedalism from experimental studies of humans and other primates." *Journal of Experimental Biology* 206, no. 9 (2003): 1437–1448.

Schultz, Adolph H. "Notes on diseases and healed fractures of wild apes and their bearing on the antiquity of pathological conditions in man." *Bulletin of the History of Medicine* 7, no. 6 (1939): 571–582.

Schwartz, Jeffrey H., and Ian Tattersall. "The human chin revisited: What is it and who has it?" *Journal of Human Evolution* 38, no. 3 (2000): 367–409.

Scott, Jeremiah E., Kevin R. McAbee, Meghan M. Eastman, and Matthew J. Ravosa. "Experimental perspective on fallback foods and dietary adaptations in early hominins." *Biology Letters* 10, no. 1 (2014a): 20130789.

Scott, Jeremiah E., Kevin R. McAbee, Meghan M. Eastman, and Matthew J. Ravosa. "Teaching an old jaw new tricks: Diet-induced plasticity in a model organism from weaning to adulthood." *Journal of Experimental Biology* 217, no. 22 (2014b): 4099–4107.

Scott, Robert S., Peter S. Ungar, Torbjorn S. Bergstrom, Christopher A. Brown, Frederick E. Grine, Mark F. Teaford, and Alan Walker. "Dental microwear texture analysis shows within-species diet variability in fossil hominins." *Nature* 436, no. 7051 (2005): 693–695.

Sears, Michael W., and Michael J. Angilletta Jr. "Costs and benefits of thermoregulation revisited: Both the heterogeneity and spatial structure of temperature drive energetic costs." *American Naturalist* 185, no. 4 (2015): E94–E102.

Seilacher, Adolf. "Arbeitskonzept zur konstruktions-morphologie." *Lethaia* 3, no. 4 (1970): 393–396.

Seilacher, Adolf. "Self-organizing mechanisms in morphogenesis and evolution." In *Constructional Morphology and Evolution*, edited by Norbert Schmidt-Kittler and Klaus Vogel, 251–271. Berlin: Springer, 1991.

Senut, Brigitte, and Christine Tardieu. "Functional aspects of Plio-Pleistocene hominid limb bones: Implications for taxonomy and phylogeny." In *Ancestors: The Hard Evidence*, edited by Eric Delson, 193–201. New York, Alan R. Liss, 1985.

Sept, Jeanne. "Modeling the significance of paleoenvironmental context for early hominin diets." In *Evolution of the Human Diet: The Known, the Unknown, and the Unknowable*, edited by Peter S. Ungar, 289–307. New York: Oxford University Press, 2007.

Seyfarth, Robert M., Dorothy L. Cheney, and Thore J. Bergman. "Primate social cognition and the origins of language." *Trends in Cognitive Sciences* 9, no. 6 (2005): 264–266.

Signor, P. W., III. "A critical evaluation of the paradigm method of functional inference." *Neues Jahrbuch für Geologie und Paläontologie-Abhandlungen* 164, no. 1–2 (1982): 59–63.

Skedros, John G. "Interpreting load history in limb-bone diaphyses: Important considerations and their biomechanical foundations." In *Bone Histology: An Anthropological Perspective*, edited by Christina Crowder and Sam Stout, 169–236. Boca Raton, FL: CRC Press, 2012.

Skedros, John G., and Sidney L. Baucom. "Mathematical analysis of trabecular 'trajectories' in apparent trajectorial structures: The unfortunate historical emphasis on the human proximal femur." *Journal of Theoretical Biology* 244, no. 1 (2007): 15–45.

Skedros, John G., Michael R. Dayton, Christian L. Sybrowsky, Roy D. Bloebaum, and Kent N. Bachus. "Are uniform regional safety factors an objective of adaptive modeling/remodeling in cortical bone?" *Journal of Experimental Biology* 206, no. 14 (2003): 2431–2439.

Skedros, John G., Kenneth J. Hunt, and Roy D. Bloebaum. "Relationships of loading history and structural and material characteristics of bone: Development of the mule deer calcaneus." *Journal of Morphology* 259, no. 3 (2004): 281–307.

Smith, Amanda L., Stefano Benazzi, Justin A. Ledogar, Kelli Tamvada, Leslie C. Pryor Smith, Gerhard W. Weber, Mark A. Spencer, et al. "The feeding biomechanics and dietary ecology of *Paranthropus boisei*." *Anatomical Record: Advances in Integrative Anatomy and Evolutionary Biology* 298, no. 1 (2015): 145–167.

Smith, Richard J. "Rethinking allometry." *Journal of Theoretical Biology* 87, no. 1 (1980): 97–111.

Smith, Richard J. "On the mechanical reduction of functional morphology." *Journal of Theoretical Biology* 96, no. 1 (1982): 99–106.

Smith, Richard J. "The mandibular corpus of female primates: Taxonomic, dietary, and allometric correlates of interspecific variations in size and shape." *American Journal of Physical Anthropology* 61, no. 3 (1983): 315–330.

Smith, Richard J. "Categories of allometry: Body size versus biomechanics." *Journal of Human Evolution* 24, no. 3 (1993): 173–182.

Smith, Richard J., and William L. Jungers. "Body mass in comparative primatology." *Journal of Human Evolution* 32, no. 6 (1997): 523–559.

Sober, Elliott. "Evolution and the problem of other minds." *Journal of Philosophy* 97, no. 7 (2000): 365–386.

Sockol, Michael D., David A. Raichlen, and Herman Pontzer. "Chimpanzee locomotor energetics and the origin of human bipedalism." *Proceedings of the National Academy of Sciences* 104, no. 30 (2007): 12265–12269.

Sorensen, Mark V., and William R. Leonard. "Neandertal energetics and foraging efficiency." *Journal of Human Evolution* 40, no. 6 (2001): 483–495.

Sponheimer, Matt, Zeresenay Alemseged, Thure E. Cerling, Frederick E. Grine, William H. Kimbel, Meave G. Leakey, Julia A. Lee-Thorp, et al. "Isotopic evidence of early hominin diets." *Proceedings of the National Academy of Sciences* 110, no. 26 (2013): 10513–10518.

Sponheimer, Matt, and Julia A. Lee-Thorp. "Isotopic evidence for the diet of an early hominid, *Australopithecus africanus*." *Science* 283, no. 5400 (1999): 368–370.

Sponheimer, Matt, Benjamin H. Passey, Darryl J. de Ruiter, Debbie Guatelli-Steinberg, Thure E. Cerling, and Julia A. Lee-Thorp. "Isotopic evidence for dietary variability in the early hominin *Paranthropus robustus*." *Science* 314, no. 5801 (2006): 980–982.

Stanford, Craig B. "The hunting ecology of wild chimpanzees: Implications for the evolutionary ecology of Pliocene hominids." *American Anthropologist* 98, no. 1 (1996): 96–113.

Sterelny, Kim. "From hominins to humans: How *sapiens* became behaviourally modern." *Philosophical Transactions of the Royal Society of London B: Biological Sciences* 366, no. 1566 (2011): 809.

Stern, Jack T., Jr. "Before bipedality." *Yearbook of Physical Anthropology* 19 (1975): 59–68.

Stern, Jack T., Jr. "The cost of bent-knee, bent-hip bipedal gait: A reply to Crompton et al." *Journal of Human Evolution* 36, no. 5 (1999): 567–570.

Stern, Jack T., Jr. "Climbing to the top: A personal memoir of *Australopithecus afarensis*." *Evolutionary Anthropology: Issues, News, and Reviews* 9, no. 3 (2000): 113–133.

Stern, Jack T., Jr., and William L. Jungers. "The capitular joint of the first rib in primates: A re-evaluation of the proposed link to locomotion." *American Journal of Physical Anthropology* 82, no. 4 (1990): 431–439.

Stern, Jack T., Jr., and Randall L. Susman. "The locomotor anatomy of *Australopithecus afarensis*." *American Journal of Physical Anthropology* 60, no. 3 (1983): 279–317.

Stern, Jack T., Jr., and Randall L. Susman. "'Total morphological pattern' versus the 'magic trait': Conflicting approaches to the study of early hominid bipedalism." *Origine(s) de la Bipédie chez les Hominidés*, edited by Yves Coppens and Brigitte Senut, 99–112. Cahiers de Paleoanthropologie Series. Paris: Éditions du Centre Nationale de la Recherche Scientifique, 1991.

Strait, David S., Paul Constantino, Peter W. Lucas, Brian G. Richmond, Mark A. Spencer, Paul C. Dechow, Callum F. Ross, et al. "Viewpoints: Diet and dietary adaptations in early hominins; The hard food perspective." *American Journal of Physical Anthropology* 151, no. 3 (2013): 339–355.

Strait, David S., Ian R. Grosse, Paul C. Dechow, Amanda L. Smith, Qian Wang, Gerhard W. Weber, Simon Neubauer, et al. "The structural rigidity of the cranium of *Australopithecus africanus*: Implications for diet, dietary adaptations, and the allometry of feeding biomechanics." *Anatomical Record: Advances in Integrative Anatomy and Evolutionary Biology* 293, no. 4 (2010): 583–593.

Strait, David S., Gerhard W. Weber, Paul Constantino, Peter W. Lucas, Brian G. Richmond, Mark A. Spencer, Paul C. Dechow, et al. "Microwear, mechanics and the feeding adaptations of *Australopithecus africanus*." *Journal of Human Evolution* 62, no. 1 (2012): 165–168.

Strait, David S., Gerhard W. Weber, Simon Neubauer, Janine Chalk, Brian G. Richmond, Peter W. Lucas, Mark A. Spencer, et al. "The feeding biomechanics and dietary ecology of *Australopithecus africanus*." *Proceedings of the National Academy of Sciences* 106, no. 7 (2009): 2124–2129, https://www.pnas.org/content/106/7/2124/.

Strait, David S., Barth W. Wright, Brian G. Richmond, Callum F. Ross, Paul C. Dechow, Mark A. Spencer, and Qian Wang. "Craniofacial strain patterns during premolar loading: Implications for human evolution." In *Primate Craniofacial Function and Biology*, edited by Christopher Vinyard, Matthew J. Ravosa, and Christine E. Wall, 173–198. Boston: Springer, 2008.

Stringer, Chris, and Peter Andrews. *The Complete World of Human Evolution*. London: Thames & Hudson, 2005.

Susman, Randall L. "Hand function and tool behavior in early hominids." *Journal of Human Evolution* 35, no. 1 (1998): 23–46.

Susman, R[andall] L., and A. B. Demes. "Relative foot length in *Australopithecus afarensis* and its implications for bipedality." *American Journal of Physical Anthropology*, Supplement 18 (1994): 192.

Susman, Randall L., and Darryl J. de Ruiter. "New hominin first metatarsal (SK 1813) from Swartkrans." *Journal of Human Evolution* 47, no. 3 (2004): 171–181.

Susman, Randall L., and Jack T. Stern Jr. "Locomotor behavior of early hominids: Epistemology and fossil evidence." *Origine(s) de la Bipédie chez les Hominidés*, edited by Yves Coppens and Brigitte Senut, 121–131. Cahiers de Paleoanthropologie Series. Paris: Éditions du Centre Nationale de la Recherche Scientifique, 1991.

Susman, Randall L., Jack T. Stern Jr., and William L. Jungers. "Arboreality and bipedality in the Hadar hominids." *Folia Primatologica* 43, no. 2–3 (1984): 113–156.

Susskind, Leonard. "Boltzmann's explanation of the second law of thermodynamics." *This Explains Everything*, edited by John Brockman, 19–21. New York: HarperCollins, 2013.

Swedell, Larissa, and Thomas Plummer. "Social evolution in Plio-Pleistocene hominins: Insights from hamadryas baboons and paleoecology." *Journal of Human Evolution* 137 (2019): 102667.

Swindler, Daris R., and Joyce E. Sirianni. "Dental size and dietary habits of primates." *Yearbook of Physical Anthropology* 19, no. 2 (1975): 167–182.

Szalay, Frederick S. "Hunting-scavenging protohominids: A model for hominid origins." *Man* 10, no. 3 (1975): 420–429.

Szalay, Frederick S., and Eric Delson. *Evolutionary History of the Primates*. New York: Academic Press, 1979.

Taleb, Nassim Nicholas. *The Black Swan: The Impact of the Highly Improbable*. New York: Random House, 2007.

Tattersall, Ian. "Species recognition in human paleontology." *Journal of Human Evolution* 15, no. 3 (1986): 165–175.

Tattersall, Ian. "Paleoanthropology: The last half-century." *Evolutionary Anthropology: Issues, News, and Reviews* 9, no. 1 (2000): 2–16.

Tattersall, Ian. "Adaptation: The unifying myth of biological anthropology." *Teaching Anthropology: Society for Anthropology in Community Colleges Notes* 9, no. 1 (2002): 9–39.

Tavella, Sara, Alessandra Ruggiu, Alessandra Giuliani, Francesco Brun, Barbara Canciani, Adrian Manescu, Katia Marozzi, et al. "Bone turnover in wild type and pleiotrophin-transgenic mice housed for three months in the International Space Station (ISS)." *PLoS One* 7, no. 3 (2012): e33179.

Taylor, Andrea B., and Dennis E. Slice. "A geometric morphometric assessment of the relationship between scapular variation and locomotion in African apes." In *Modern Morphometrics in Physical Anthropology*, edited by Dennis E. Slice, 299–318. Boston: Springer, 2006.

Taylor, Andrea B., Tian Yuan, Callum F. Ross, and Christopher J. Vinyard. "Jaw-muscle force and excursion scale with negative allometry in platyrrhine primates." *American Journal of Physical Anthropology* 158, no. 2 (2015): 242–256.

Taylor, C. Richard, and V. J. Rowntree. "Running on two or on four legs: Which consumes more energy?" *Science* 179, no. 4069 (1973): 186–187.

Teaford, Mark F., and Alan Walker. "Quantitative differences in dental microwear between primate species with different diets and a comment on the presumed diet of *Sivapithecus*." *American Journal of Physical Anthropology* 64, no. 2 (1984): 191–200.

Thayer, Zaneta M., and Seth D. Dobson. "Sexual dimorphism in chin shape: Implications for adaptive hypotheses." *American Journal of Physical Anthropology* 143, no. 3 (2010): 417–425.

Thomas, Roger D. K., and W.-E. Reif. "The skeleton space: A finite set of organic designs." *Evolution* 47, no. 2 (1993): 341–360.

Tobias, Phillip V. "Single characters and the total morphological pattern redefined: The sorting effected by a selection of morphological features of the early hominids." In *Ancestors: The Hard Evidence*, edited by Eric Delson, 94–101. New York: Alan R. Liss, 1985.

Trichilo, Silvia, and Peter Pivonka. "Application of disease system analysis to osteoporosis: From temporal to spatio-temporal assessment of disease progression and intervention." In *Multiscale Mechanobiology of Bone Remodeling and Adaptation*, 61–121. Cham, Switzerland: Springer, 2018.

Trinkaus, Erik. 1989. "The Upper Pleistocene transition." In *The Emergence of Modern Humans: Biocultural Adaptations in the Later Pleistocene*, edited by Erik Trinkaus, 42–66. Cambridge: Cambridge University Press, 1989.

Truzzi, Marcello. "Judging the Hill Case." In *Encounters at Indian Head: The Betty & Barney Hill UFO Abduction Revisited*, edited by Karl T. Pflock and Peter Brookesmith, 70–89. San Antonio: Anomalist Books, 2007.

Turner, Charles H., Ichiro Owan, and Yuichi Takano. "Mechanotransduction in bone: Role of strain rate." *American Journal of Physiology: Endocrinology and Metabolism* 269, no. 3 (1995): E438–E442.

Turner, Charles H., and Alexander G. Robling. "Designing exercise regimens to increase bone strength." *Exercise and Sport Sciences Reviews* 31, no. 1 (2003): 45–50.

Tuttle, Russell H. "Knuckle-walking and the problem of human origins." *Science* 166, no. 3908 (1969): 953–961.

Tuttle, Russell H. "Evolution of hominid bipedalism and prehensile capabilities." *Philosophical Transactions of the Royal Society of London: Series B, Biological Sciences* 292, no. 1057 (1981): 89–94.

Tuttle, Russell H. "Ape footprints and Laetoli impressions: A response to SUNY claims." In *Hominid Evolution: Past, Present, and Future*, edited by Phillip V. Tobias, 129–133. New York: Alan R. Liss, 1985.

Tuttle, Russel H., David M. Webb, and Michael Baksh. "Laetoli toes and *Australopithecus afarensis*." *Human Evolution* 6, no. 3 (1991): 193–200.

Ungar, Peter [S.]. "Dental allometry, morphology, and wear as evidence for diet in fossil primates." *Evolutionary Anthropology: Issues, News, and Reviews* 6, no. 6 (1998): 205–217.

Ungar, Peter S., ed. *Evolution of the Human Diet: The Known, the Unknown, and the Unknowable*. New York: Oxford University Press, 2006.

Ungar, Peter S., Frederick E. Grine, and Mark F. Teaford. "Dental microwear and diet of the Plio-Pleistocene hominin *Paranthropus boisei*." *PLoS One* 3, no. 4 (2008): e2044.

Ungar, Peter S., and Leslea J. Hlusko. "The evolutionary path of least resistance." *Science* 353, no. 6294 (2016): 29–30.

Ungar, Peter S., and Francis M'Kirera. "A solution to the worn tooth conundrum in primate functional anatomy." *Proceedings of the National Academy of Sciences* 100, no. 7 (2003): 3874–3877.

Ungar, Peter S., Robert S. Scott, Frederick E. Grine, and Mark F. Teaford. "Molar microwear textures and the diets of *Australopithecus anamensis* and *Australopithecus afarensis*." *Philosophical Transactions of the Royal Society B: Biological Sciences* 365, no. 1556 (2010): 3345–3354.

Ungar, Peter S., Jessica R. Scott, and Christine M. Steininger. "Dental microwear differences between eastern and southern African fossil bovids and hominins." *South African Journal of Science* 112, no. 3–4 (2016): 1–5.

Ungar, P[eter] S., and Malcolm Williamson. "Exploring the effects of tooth wear on functional morphology: A preliminary study using dental topographic analysis." *Palaeontologia Electronica* 3, no. 1 (2000): 1–18.

van Der Merwe, Nikolaas J., J. Francis Thackeray, Julia A. Lee-Thorp, and Julie Luyt. "The carbon isotope ecology and diet of *Australopithecus africanus* at Sterkfontein, South Africa." *Journal of Human Evolution* 44, no. 5 (2003): 581–597.

van Der Meulen, M. C. H., G. S. Beaupré, and D. R. Carter. "Mechanobiologic influences in long bone cross-sectional growth." *Bone* 14, no. 4 (1993): 635–642.

Van Valen, Leigh. "Morphological variation and width of ecological niche." *American Naturalist* 99, no. 908 (1965): 377–390.

Vashishth, D., O. Verborgt, G. Divine, M. B. Schaffler, and D. P. Fyhrie. "Decline in osteocyte lacunar density in human cortical bone is associated with accumulation of microcracks with age." *Bone* 26, no. 4 (2000): 375–380.

Via, Sara. "Genetic constraints on the evolution of phenotypic plasticity." In *Genetic Constraints on Adaptive Evolution*, edited by Volker Loeschcke, 47–71. Berlin: Springer, 1987.

Via, Sara, and Russell Lande. "Genotype-environment interaction and the evolution of phenotypic plasticity." *Evolution* 39, no. 3 (1985): 505–522.

Villmoare, Brian. "Morphological integration, evolutionary constraints, and extinction: A computer simulation–based study." *Evolutionary Biology* 40, no. 1 (2013): 76–83.

Vinyard, Christopher J., Christine E. Wall, Susan H. Williams, and William L. Hylander. "Patterns of variation across primates in jaw-muscle electromyography during mastication." *American Zoologist* 48, no. 2 (2008): 294–311.

Vogel, Klaus. "Concepts of constructional morphology." In *Constructional Morphology and Evolution*, edited by Norbert Schmidt-Kittler and Klaus Vogel, 55–68. Berlin: Springer, 1991.

Vogel, Steven. *Life's Devices: The Physical World of Animals and Plants*. Princeton, NJ: Princeton University Press, 1988.

Vrba, Elisabeth S. "Evolution, species and fossils: How does life evolve?" *South African Journal of Science* 76, no. 2 (1980): 61–84.

Vrba, Elisabeth S. "Environment and evolution: Alternative causes of the temporal distribution of evolutionary events." *South African Journal of Science* 81, no. 5 (1985): 229–236.

Waddington, C. H. "Canalization of development and the inheritance of acquired characters." *Nature* 150, no. 3811 (1942): 563–565.

Wagner, Günter P. "What is the promise of developmental evolution? Part I, why is developmental biology necessary to explain evolutionary innovations?" *Journal of Experimental Zoology* 288, no. 2 (2000): 95–98.

Wagner, Günter P., and Lee Altenberg. "Perspective: Complex adaptations and the evolution of evolvability." *Evolution* 50, no. 3 (1996): 967–976.

Wagner, G[ünter] P. and B. Y. Misof. "How can a character be developmentally constrained despite variation in developmental pathways?" *Journal of Evolutionary Biology* 6, no. 3 (1993): 449–455.

Wake, David B. "Homoplasy: The result of natural selection, or evidence of design limitations?" *American Naturalist* 138, no. 3 (1991): 543–567.

Wainwright, Peter C., and Stephen M. Reilly, eds. *Ecological Morphology: Integrative Organismal Biology.* Chicago: University of Chicago Press, 1994.

Wainwright, Stephen A. "Form and function in organisms." *American Zoologist* 28, no. 2 (1988): 671–680.

Walker, Alan, Hendrick N. Hoeck, and Linda Perez. "Microwear of mammalian teeth as an indicator of diet." *Science* 201, no. 4359 (1978): 908–910.

Walker, Christopher S., and Steven E. Churchill. "Territory size in *Canis lupus*: Implications for Neandertal mobility." In *Reconstructing Mobility*, edited by Kristian J. Carlson and Damiano Marchi, 209–226. Boston: Springer, 2014.

Wallace, Ian J., Brigitte Demes, Carrie Mongle, Osbjorn M. Pearson, John D. Polk, and Daniel E. Lieberman. "Exercise-induced bone formation is poorly linked to local strain magnitude in the sheep tibia." *PLoS One* 9, no. 6 (2014): e99108.

Wallace, Ian J., Steven M. Tommasini, Stefan Judex, Theodore Garland Jr., and Brigitte Demes. "Genetic variations and physical activity as determinants of limb bone morphology: An experimental approach using a mouse model." *American Journal of Physical Anthropology* 148, no. 1 (2012): 24–35.

Wallace, John A. "Dietary adaptations of *Australopithecus* and early *Homo*." In *Paleoanthropology: Morphology and Paleoecology*, edited by Russell H. Tuttle, 203–223. The Hague: Mouton, 1975.

Wang, Qian, Amanda L. Smith, David S. Strait, Barth W. Wright, Brian G. Richmond, Ian R. Grosse, Craig D. Byron, and Uriel Zapata. "The global impact of sutures assessed in a finite element model of a macaque cranium." *Anatomical Record: Advances in Integrative Anatomy and Evolutionary Biology* 293, no. 9 (2010): 1477–1491.

Ward, Carol V. "Interpreting the posture and locomotion of *Australopithecus afarensis*: Where do we stand?" *American Journal of Physical Anthropology* 119, no. S35 (2002): 185–215.

Warden, Stuart J., and Robyn K. Fuchs. "Exercise and bone health: Optimising bone structure during growth is key, but all is not in vain during ageing." *British Journal of Sports Medicine* 43, no. 12 (2009): 885–887.

Weber, Hermann. *Stellung und Aufgaben der Morphologie in der Zoologie der Gegenwart.* Leipzig: Geest & Portig, 1954.

Webster, G., and B. C. Goodwin. "The origin of species—a structuralist approach." *Journal of Social and Biological Structures* 5, no. 1 (1982): 15–47.

Weidenreich, Franz. "The mandibles of *Sinanthropus pekinensis*: A comparative study." *Paleontologia Sinica*, ser. D, 7, no. 3 (1936): 1–162.

West-Eberhard, Mary Jane. "Phenotypic accommodation: Adaptive innovation due to developmental plasticity." *Journal of Experimental Zoology, Part B: Molecular and Developmental Evolution* 304, no. 6 (2005): 610–618.

Westerlind, Kim C., Thomas J. Wronski, Erik L. Ritman, Zong-Ping Luo, Kai-Nan An, Norman H. Bell, and Russell T. Turner. "Estrogen regulates the rate of bone turnover but bone balance in ovariectomized rats is modulated by prevailing mechanical strain." *Proceedings of the National Academy of Sciences* 94, no. 8 (1997): 4199–4204.

Westerterp, Klaas R., Gerwin A. L. Meijer, Eugene M. E. Janssen, Wim H. M. Saris, and Foppe Ten Hoor. "Long-term effect of physical activity on energy balance and body composition." *British Journal of Nutrition* 68, no. 1 (1992): 21–30.

Wheeler, Peter E. "The thermoregulatory advantages of hominid bipedalism in open equatorial environments: The contribution of increased convective heat loss and cutaneous evaporative cooling." *Journal of Human Evolution* 21, no. 2 (1991): 107–115.

White, Tim D. "A view on the science: Physical anthropology at the millennium." *American Journal of Physical Anthropology* 113, no. 3 (2000): 287–292.

White, Tim D., Berhane Asfaw, Yonas Beyene, Yohannes Haile-Selassie, C. Owen Lovejoy, Gen Suwa, and Giday WoldeGabriel. "*Ardipithecus ramidus* and the paleobiology of early hominids." *Science* 326, no. 5949 (2009): 64–86.

Whiteside, Mark A., Rufus Sage, and Joah R. Madden. "Multiple behavioural, morphological and cognitive developmental changes arise from a single alteration to early life spatial environment, resulting in fitness consequences for released pheasants." *Royal Society Open Science* 3, no. 3 (2016): 160008.

Williams, George C. *Adaptation and Natural Selection.* Princeton, NJ: Princeton University Press, 1966.

Williams, George C. *Natural Selection: Domains, Levels, and Challenges.* New York: Oxford University Press, 1992.

Witmer, Lawrence M. "The extant phylogenetic bracket and the importance of reconstructing soft tissues in fossils." In *Functional Morphology in Vertebrate Paleontology*, edited by J. J. Thomason, 19–33. Cambridge: Cambridge University Press, 1995.

Wolpoff, Miford H. "Some aspects of human mandibular evolution." In *Determinants of Mandibular Growth and Form*, edited by James A. McNamara Jr., 1–64. Ann Arbor: Center for Human Growth and Development, University of Michigan, 1975.

Wolpoff, Milford H. *Paleoanthropology.* New York: Alfred A. Knopf, 1980.

Wolpoff, Milford H. "Lucy's lower limbs: Long enough for Lucy to be fully bipedal?" *Nature* 304, no. 5921 (1983): 59–61.

Wood, Bernard. "Early hominid species and speciation." *Journal of Human Evolution* 22, no. 4–5 (1992): 351–365.

Wood, Bernard, and Mark Collard. "The changing face of genus *Homo*." *Evolutionary Anthropology: Issues, News, and Reviews* 8, no. 6 (1999): 195–207.

Wood, Bernard, and Daniel E. Lieberman. "Craniodental variation in *Paranthropus boisei*: A developmental and functional perspective." *American Journal of Physical Anthropology* 116, no. 1 (2001): 13–25.

Wood, Bernard, and David Strait. "Patterns of resource use in early *Homo* and *Paranthropus*." *Journal of Human Evolution* 46, no. 2 (2004): 119–162.

Wroe, Stephen, Toni L. Ferrara, Colin R. McHenry, Darren Curnoe, and Uphar Chamoli. "The craniomandibular mechanics of being human." *Proceedings of the Royal Society B: Biological Sciences* 277, no. 1700 (2010): 3579–3586.

Wynn, Jonathan G., Matt Sponheimer, William H. Kimbel, Zeresenay Alemseged, Kaye Reed, Zelalem K. Bedaso, and Jessica N. Wilson. "Diet of *Australopithecus afarensis* from the Pliocene Hadar Formation, Ethiopia." *Proceedings of the National Academy of Sciences* 110, no. 26 (2013): 10495–10500.

Young, Nathan M. "Modularity and integration in the hominoid scapula." *Journal of Experimental Zoology, Part B: Molecular and Developmental Evolution* 302, no. 3 (2004): 226–240.

Young, Nathan M., Günter P. Wagner, and Benedikt Hallgrímsson. "Development and the evolvability of human limbs." *Proceedings of the National Academy of Sciences* 107, no. 8 (2010): 3400–3405.

Zihlman, Adrienne L., John E. Cronin, Douglas L. Cramer, and Vincent M. Sarich. "Pygmy chimpanzee as a possible prototype for the common ancestor of humans, chimpanzees and gorillas." *Nature* 275, no. 5682 (1978): 744–746.

Index